Advances in STEM Education

Advances in STEM Education is a new book series with a focus on cutting-edge research and knowledge development in science, technology, engineering and mathematics (STEM) education from pre-college through continuing education around the world. It is open to all topics in STEM education, both in and outside of classrooms, including innovative approaches and perspectives in promoting and improving STEM education, and the processes of STEM instruction and teacher education. This series values original contributions that view STEM education either in terms of traditionally defined subject-based education or as an educational undertaking involving inter-connected STEM fields. The series is open to new topics identified and proposed by researchers internationally, and also features volumes from invited contributors and editors. It works closely with the*International Journal of STEM Education* and*Journal for STEM Education Research* to publish volumes on topics of interest identified from the journal publications that call for extensive and in-depth scholarly pursuit.

Researchers interested in submitting a book proposal should contact the Series Editor: Yeping Li (yepingli@tamu.edu) or the Publishing Editor: Melissa James (Melissa.James@springer.com) for further information.

More information about this series at http://www.springer.com/series/13546

Stephen Miles Uzzo • Sherryl Browne Graves
Erin Shay • Marisa Harford • Robert Thompson
Editors

Pedagogical Content Knowledge in STEM

Research to Practice

 Springer

Editors
Stephen Miles Uzzo
New York Hall of Science
Corona, NY, USA

Erin Shay
School of Education
Hunter College
New York, NY, USA

Robert Thompson
Department of Mathematics and Statistics
Hunter College
New York, NY, USA

Sherryl Browne Graves
Hunter College
City University of New York
New York City, NY, USA

Marisa Harford
New Visions for Public Schools
New York, NY, USA

ISSN 2520-8616   ISSN 2520-8624   (electronic)
Advances in STEM Education
ISBN 978-3-030-07361-9   ISBN 978-3-319-97475-0   (eBook)
https://doi.org/10.1007/978-3-319-97475-0

This Springer imprint is published by the registered company Springer Nature Switzerland AG
The registered company address is: Gewerbestrasse 11, 6330 Cham, Switzerland

Preface

Thomas Kuhn famously compared scientific paradigms to political ones in that there are communities of theory and practice that must be in fundamental agreement for such paradigms to flourish: "this issue of paradigm choice can never be unequivocally settled by logic and experiment alone." In the twenty-first century, we recognize that these agreements are, just as science is, transitory: true until disproven or made obsolete. Science, by its very process, is an evolving set of paradigms for investigating the phenomena of nature, including new ones that emerge and suddenly (in a matter of decades) change everything. This creates a terrible mess for educators. How does one teach a science in constant flux? How do we reconcile the core domains of science (such as chemistry, physics, and biology), when they look at the behavior and substance of matter very differently, at least for now?

Of course, this dilemma is not limited to the pure sciences, but also the soft sciences, engineering, math, and the evolving technological tools for doing science and engineering, what we commonly call "STEM, as well as virtually all other domains of knowledge." In 1984, Lee Shulman saw this challenge as an opportunity to explicitly embrace the problem through specifically targeting domains and topics with clear pathways into pedagogy for each. For the sake of this volume, there is no one-size-fits-all teaching or learning approach to all of STEM. Shulman coined the term *Pedagogical Content Knowledge* (PCK) to indicate that, in order to effectively guide the learning process, practitioners must employ teaching approaches that are particular to the content knowledge for the topic and domain in which they are teaching and that there are, perhaps different ways into deepening engagement with, for instance, redox reactions in chemistry, than there is with ATP synthesis in biology.

PCK has evolved into a diverse area of learning and education research with the potential to transform teaching and learning throughout STEM domains and across learning settings. This volume brings together experts in pedagogical content knowledge to describe cases of bringing PCK into a wide variety of learning settings and the results of those efforts. Some authors chose to present cases describing the structure and process of such deployment, while others describe specific approaches to research PCK in live formal and informal learning environments, and others describe barriers and opportunities for bringing PCK research into practice.

The goal of this work is to give the reader and practitioner a well-rounded sense of the gains being made in STEM teaching and learning through PCK, how the research is informing educational practice, and ultimately inspire practitioners into bringing these ideas into their own thinking about STEM learning.

A variety of approaches to the task are presented in this volume, including descriptions of cases to engage learners in enhanced PCK instruction, research of PCK in learning environments, and addressing the challenges and opportunities of moving PCK from research into practice. Chapter 1 provides an extensive introduction to PCK including a consensus definition. They then provide examples of two methods for measuring teacher PCK. Chapter 2 draws upon Content Knowledge for Teaching (CKT) and addresses means of assessing Disciplinary CKT and Pedagogical CKT. Authors describe a method of measuring CKT-D and CKT-P and their support of student thinking. Chapter 3 addresses the synergy of Personal PCK and Canonical PCK, introducing the possibility of this synergy. Future research as described by the authors may help to contribute more to the field. Chapter 4 introduces barriers and supports for science instruction as described by participants of a professional development program that addresses the development of PCK for the STEM educator. Chapter 5 extends the discussion of PCK to include Mathematics Knowledge for Teaching specific to Visual Representations (MKT-VR). The chapter provides a sound description of MKT-VR and method for measuring the MKT-VR of STEM educators. Chapter 6 looks closely at the role of teacher inquiry in PCK development. Chapter 7 focuses on how to modify a graduate school biology syllabus to advantage PCK. Chapter 8 introduces the reader into the modes of PCK integration across several mathematics courses for students in the MASTER residency program. Chapter 9 provides an overview of how PCK is addressed and measured in the project and the changes occurred across iterations. The MASTER program itself presents an innovative method of preparing preservice STEM educators and providing a variety of contextual learning experiences that support the development of PCK in STEM education. Chapters 10 and 11 describe how experiences in an informal environment can support the development of PCK for teaching inquiry and integration of engineering into mathematics of science instruction. The small pilot group gave indications that the practices described may support the development of inquiry teaching PCK in preservice teachers. In both Chaps 10 and 11, the authors found that the explicit scaffolds offered through the museum partnership improved the practice of participants during the experience. They suggest looking further into how participants integrated the modeling and engineering into their classroom teaching. Chapter 12 looks at the role of an informal learning institution in a PCK-based collaborative residency program. Chapter 13 addresses the practice of argumentation as a piece of the PCK required by STEM educators. In their chapter, they describe the potential impacts of educative curriculum materials (ECM) on the development of PCK with regard to argumentation. In particular, they investigate the distinction between ECM that includes multimedia elements and how they support PCK. Chapter 14 introduces the realm of educative making into PCK. The authors argue that as more maker spaces find their way into schools and other learning environments, the PCK of teachers in educative making moments needs to be further researched and in turn developed. They propose teacher education practice to support a model of maker pedagogical content knowledge (MPCK).

We hope the voices of the many authors represented in this volume will stimulate needed conversation about the challenges in implementing and scaling PCK in the clinic and the role of learning institutions of all kinds in supporting this important idea throughout lifelong learning. While this book represents many dimensions of PCK in STEM, it is merely a point in time, and there is more to discover and new challenges to education to address as interdisciplinary and data-driven science practices and processes find their way more deeply into teaching and learning. Of course, PCK will evolve with it and, we believe, will be as relevant in the twenty-second century as it is for the twenty-first.

Corona, NY, USA   Stephen Miles Uzzo
New York, NY, USA   Sherryl Browne Graves
                                                                                              Erin Shay
                                                                                         Marisa Harford
                                                                                     Robert Thompson

Contents

About the Editors

Stephen Miles Uzzo is chief scientist for the New York Hall of Science where he does research and development of public programs and experiences on complex science and systems dynamics. He also does instructional development for preservice and in-service teacher education. Dr. Uzzo is adjunct professor of education for the New York Institute of Technology Graduate School of Education and Interdisciplinary Studies, where he teaches STEM integration of science, technology, engineering, and mathematics into science instruction, instructional technology, and integration of art into interdisciplinary teaching. His background includes teaching and learning in data-driven science, computer graphics systems engineering, environmental science, and art history. He holds a doctorate of network theory and environmental studies from the Union Institute School of Interdisciplinary Studies and a bachelor of education from Long Island University Department of Music.

Sherryl Browne Graves is the acting senior associate dean of education at Hunter College and teaches courses in psychological foundations of education including courses in child development, educational psychology, educational research, cognition and educational technology, and multicultural issues in learning and instruction. Professor Graves' research interests focus on children's understanding of racial and ethnic portrayals in mass media, the effects of diversity in the educational process, and the use of technology in teaching and learning. She has served as a consultant and advisory board member to numerous media organizations including Sesame Workshop, WGBH and KCET Public Television Stations, Discovery Kids, and the Public Broadcasting Service. Additionally she has served as a PI on an NSF MSP grant, co-PI on a New York Community Trust early literacy grant, and senior advisor on a Foundation for Child Development PreK for All grant. Dr. Graves is trained in psychology with a doctorate from Harvard University in clinical psychology and public practice and a bachelor's degree in psychology from Swarthmore College.

Erin Shay is the director of the Office of Partnership Programs at Hunter College where she oversees the School of Education's alternative certification and partnership programs. Working in collaboration with a variety of local and national organizations, the Office of Partnership Programs offers a wide range of teacher preparation programs, additional certificate programs for currently certified teachers to become certified in a second specialization area, and educational leadership programs to prepare teachers for roles in administration and supervision. She holds a bachelor's degree in education and social policy from Northwestern University and a master's degree in education from Bank Street College of Education. Ms. Shay is a former NYC public school math and English teacher.

Marisa Harford is the director of teacher residencies at New Visions for Public Schools, a nonprofit organization that partners with New York City schools, communities, and institutions to strengthen schools through multiple channels, including developing and retaining effective teachers through innovative residency programs in partnership with Hunter College and the NYC Department of Education. Ms. Harford leads the teacher certification team, which prepares highly qualified novice teachers, builds capacity in mentoring and support of new teacher development, and crafts sustainable systems and tools for residencies. Her previous experience includes 7 years of secondary-level teaching and instructional coaching at Bronx middle and high schools. She holds a BA in English from Yale University, an MS in secondary English education from Lehman College, and an advanced certificate in school and building leadership from Hunter College.

Robert Thompson is a professor of mathematics at Hunter College, CUNY, and the CUNY Graduate Center. He was chair of the Department of Mathematics and Statistics from 2010 to 2017. Professor Thompson's research interests are in algebraic topology and homotopy theory. He teaches courses at all levels, from introductory college mathematics courses for liberal arts majors to graduate courses in mathematics for masters' and PhD students. In addition to topology, Professor Thompson has worked with students and colleagues on projects in probability theory, mathematical finance, applications of mathematics to biology and robotics, and teacher preparation. Professor Thompson has a doctorate in mathematics from the University of Washington and bachelor's degree in mathematics from the University of California at Santa Cruz.

Part I
PCK Research in Formal Teaching Practice

Chapter 1
Analysis of Practice and Teacher PCK: Inferences from Professional Development Research

Christopher D. Wilson, Molly Stuhlsatz, Connie Hvidsten, and April Gardner

Abstract In this chapter, we describe an approach to professional development grounded in a model of PCK that centers around a teacher's knowledge of how to elicit and respond to student thinking, to engage students in the practices of science, and to develop coherent and effective learning experiences. This approach has been shown to be effective in increasing PCK in both in-service professional development and pre-service teacher education contexts and has resulted in increased student achievement when compared to more traditional approaches. Current and future directions for this work include examining the pathways between teacher professional knowledge and student learning, further development of our approach to measuring PCK using automated analysis of teacher writing, and exploring how to effectively scale up this somewhat resource-intensive approach to teacher learning.

Keywords PCK · Professional development · Lesson analysis · Student learning · Personal PCK and skill · Video analysis · Analysis of practice · Student content knowledge

1.1 Introduction

In this chapter, we describe a professional development approach for developing PCK that involves teachers engaged in analyzing teaching practice using video recordings of classroom sessions. This approach also includes identifying and using instructional strategies that have been shown effective in enhancing student learning.

C. D. Wilson (✉) · M. Stuhlsatz · C. Hvidsten · A. Gardner
BSCS Science Learning, Colorado Springs, CO, USA

© Springer International Publishing AG, part of Springer Nature 2018
S. M. Uzzo et al. (eds.), *Pedagogical Content Knowledge in STEM*,
Advances in STEM Education, https://doi.org/10.1007/978-3-319-97475-0_1

1.2 The Introduction of PCK

Studies showing weak or inconclusive links between teachers' content knowledge and student achievement have puzzled researchers for nearly 40 years. In considering the dilemma, Lee Shulman and colleagues rejected the description of teachers "as mere emitters of behavior" and became more concerned with "the motives and implicit reasoning that explained teacher and student behavior" (Shulman 2002). They proposed a "missing paradigm" in educational research in 1986, with the idea of pedagogical content knowledge (PCK). PCK challenged past practices of examining knowledge of subject matter and pedagogy separately. Instead, PCK recognizes the melding of subject matter expertise with pedagogical strategies and knowledge of the learner to produce high-quality classroom practice. For Shulman and the researchers that followed, PCK is a unique teacher knowledge base that allows them to consider the structure and importance of an instructional topic, recognize the features that will make it more or less accessible to students, and justify the selection of teaching practices based on student learning needs.

This idea clarified and generated new ways of thinking about the relationships between teacher knowledge and teacher practice (Shulman 1987). Shulman's description of PCK resonated with many educational researchers and initiated much research on professional teacher knowledge. Over the next 25 years, science education researchers enthusiastically studied PCK (Abell 2008; Borko and Livingston 1989; Cochran et al. 1993; Gess-Newsome 1999; Grossman 1990; Hashweh 1987; Heller et al. 2004; Kind 2009; Loughran et al. 2001; Magnusson et al. 1999; Padilla et al. 2008; Park and Oliver 2008; Rollnick et al. 2008; Schneider and Plasman 2011; Van Driel et al. 1998). However, because the PCK construct was based on a conceptualization rather than empirical study, researchers were free to reconceptualize the idea of PCK. The outcome of this array of research over the ensuing years was divergence in the interpretation and understanding of PCK. In an address to PCK researchers in 2012, Shulman himself (2015) noted:

> In some ways I feel a little uncomfortable talking authoritatively about pedagogical content knowledge now. I feel like the biological father of a baby that was raised in its infancy and then given away for adoption or foster care when it was about five years old. During the years that followed, the youngster was raised by many parents and played with many peers. Now that it has survived adolescence and reached emergent adulthood, most of you know far more about PCK that I possibly could because you have been living with, developing, elaborating, revising, and applying that set of ideas in serious research and pedagogical work.

Following Shulman's introduction of the PCK construct, other researchers included pedagogical content knowledge as one of the knowledge bases for teaching. The most complete descriptions from these early years consisted of four components: general pedagogical knowledge, subject matter knowledge, PCK, and knowledge of context (Grossman 1990). Magnusson et al. (1999) started with this description to develop a model of PCK for science teaching. Their model describes five components of PCK: orientations to teaching science, knowledge of science curricula, knowledge of assessment of scientific literacy, knowledge of students' understanding of science, and knowledge of instructional strategies. Many subsequent science education research studies have used the Magnusson model as originally described or with modifications based on the perspectives and findings of the studies, including making teaching orientation a backdrop rather than a component of PCK (Friedrichesen et al. 2012). This shift gave more emphasis to the relationships among the components in the Magnusson model (Park and Oliver 2008; van Driel and Henze 2012) and adding knowledge of how to sequence ideas (Smith and Banilower 2012). Other researchers have described PCK for science teaching as an integration of content knowledge, pedagogical knowledge, and contextual knowledge (Gess-Newsome 1999), PCK as completely distinct from pedagogical knowledge and content knowledge (Kirschner et al. 2012), or PCK as a transformation of knowledge of student context, knowledge of student ideas about concepts, subject matter knowledge, and content knowledge (Rollnick and Mavhunga 2012).

1.3   PCK Summits and Consensus Definition

The growing divergence among the models of PCK used by science education researchers was the impetus for the first "PCK Summit" held in the fall of 2012 (Carlson et al. 2015). Researchers of PCK in science and mathematics teaching gathered to compare and negotiate their differing models, with the goal of developing a consensus model of PCK. Despite the divergence of PCK models used by these researchers, they were able to identify commonalities among them. All included content or subject matter knowledge and aspects of pedagogical knowledge, such as knowledge of instructional strategies, assessment, or curricula. Many also included understanding of the school or student context. Among science education researchers, there was general agreement that PCK is topic-specific, rather than domain-specific. That is, there is not a general PCK for teaching physics, biology, or chemistry, but PCK for teaching acceleration, photosynthesis, or covalent bonds.

One confounding issue for participants was whether PCK is a knowledge base that can be assessed through paper-and-pencil tests or interviews, or a skill that can be assessed only through classroom observations, or both. Another problematic issue was whether PCK is canonical – a knowledge base that can be taught – or personal and must be gained through experience and reflection. The PCK Summit participants agreed on the following definition of personal PCK and PCK and skill (Gess-Newsome 2015):

- Personal PCK is the knowledge of, reasoning behind, and planning for teaching a particular topic in a particular way for a particular purpose to particular students for enhanced student outcomes.
- Personal PCK and skill is the act of teaching a particular topic in a particular way for a particular purpose to particular students for enhanced student outcomes.

The consensus model resulting from the Summit includes PCK as one of several knowledge bases within teacher professional knowledge and skill. It includes both topic-specific professional knowledge (which might be considered a canonical form of PCK) and personal PCK, which is developed through experience and described above. As the name "teacher professional knowledge and skill" implies, the model includes knowledge that is held by an individual as well as skill in enacting that knowledge. See Gess-Newsome (2015) for a detailed description of the components of this model and the relationships among them.

1.4 Approaches to Enhancing PCK Among Teachers

Loughran et al. (2012) developed Content Representations (CoRes) as a tool for capturing teachers' PCK. The CoRes scaffolds and documents teachers' thinking about specific science topics by asking them to identify the "big ideas" associated with a topic and responding to several prompts about each of these concepts, such as what students should learn related to a big idea, why they should learn that, what students typically find challenging about the concept, and teaching strategies that are effective for helping students learn. Researchers who have used CoRes found the tool to be useful not only for collecting examples of expert PCK but also as a means of enhancing or accelerating the development of more sophisticated PCK among beginning science teachers (Bertram 2014; Hume 2010; Williams 2012). Other approaches for enhancing PCK among science teachers involve specially designed professional development programs. For example, Gess-Newsome et al. (2017) designed a 2-year, curriculum-based professional development intervention that successfully increased biology teachers' PCK. Daehler et al. (2015) described the development of a science teacher professional development program to develop PCK that includes science investigations coupled with examination of teaching cases and reflections about connections to their students and classrooms.

1.5 Science Teachers Learning Through Lesson Analysis (STeLLA)

Science Teachers Learning Through Lesson Analysis (STeLLA) is a series of professional development and teacher preparation programs grounded in research on how teachers learn and built around core science teaching practices delineated in the STeLLA conceptual framework (Roth et al. 2011; Taylor et al. 2016). Initially, STeLLA described a 1-year professional development program for upper elementary teachers (STeLLA 1 was an initial test of the efficacy of the approach, whereas STeLLA 2 tested the approach in a randomized controlled experiment described later in this chapter). The STeLLA approach has now been scaled up and is used in both in-service and pre-service settings for teachers of kindergarten through 12th grade, in face-to-face, online, and leadership development contexts.

The essential features of the STeLLA approach that lead to enhanced teacher PCK include (1) the goals for teacher learning that intertwine increasing pedagogical knowledge with increasing science content knowledge, (2) the structure of the program that scaffolds teacher learning in ways that lead from greater support for teachers to greater independence and accountability, and (3) the context of teacher learning featuring analysis of grade- and content-specific classroom videos.

1.5.1 STeLLA Teacher Learning Goals

In programs using the STeLLA approach, teachers learn to analyze science teaching through two lenses – a "Student Thinking Lens" that focuses teachers' attention on ways to reveal, support, and challenge student thinking and a "Science Content Storyline Lens" that focuses teachers' attention on the coherence and connections that students can build as they learn science (Fig. 1.1). These two lenses are instantiated through 18 strategies that support teachers as they develop student thinking and coherent storyline-centered approaches to teaching science in their particular context.

In addition to the STeLLA lenses and strategies, the STeLLA approach features science content learning goals for teachers that focus on a deep understanding of two science content areas taught at the grade level they teach. Teachers deepen their understanding of content through activities and experiences that model teaching and learning of science content using the STeLLA lenses and strategies. For example, teachers in a summer institute may participate as learners in a science activity led by PD leaders who model the use of the STeLLA strategies by asking questions that elicit, probe, or challenge teacher thinking (STeLLA strategies 1, 2, and 3). They engage in analyzing data, using models, or developing explanations and arguments about phenomena related to their grade-level standards (STeLLA strategies 5, 6, and 7). The sequence of science activities they encounter will reflect a carefully designed conceptual flow (STeLLA Science Content Storyline strategies) with opportunities for teachers to make links between science ideas and the science content-based activities they complete. The activities teachers engage in during a summer institute are similar to activities they will lead with their own students, but designed to deepen their understanding of the science as adult learners. Teachers further deepen their understanding of these science concepts as they study the STeLLA lesson plans provided for their teaching and the common ideas and unscientific ways of thinking they might encounter when teaching this content to their students. In addition, teachers collaboratively analyze video of students discussing their ideas related to the science content or reflect on authentic student artifacts. Each of these experiences provides an opportunity for teachers to learn how to effectively use teaching strategies in ways that enhance teaching of specific science content within a particular classroom situation and to transfer this knowledge into their own teaching practice.

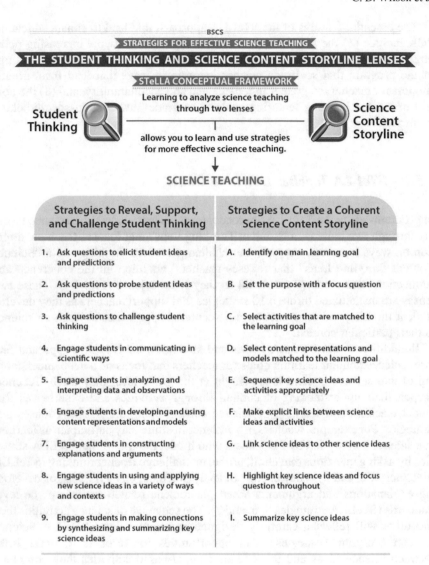

Fig. 1.1 The STeLLA lenses and strategies

1.5.2 STeLLA Program Structure

The structure of the STeLLA approach is grounded in situated cognition learning theory and a cognitive apprenticeship model of instruction (Collins 2006; Lave and Wenger 1999). Teacher learning is situated in the contexts that the knowledge will be used – classrooms. Each aspect of the teacher learning experience is centered on experiences that enhance their classroom teaching: analyzing classroom video (discussed in the next section), studying lesson plans or student artifacts, engaging in science activities, or reflecting on their own teaching experiences. The program is structured to begin with learning about science content and effective pedagogy from those with expertise in the STeLLA approach and continues with experiences that slowly fade the support of experts and turn greater independence of use to the participating teachers. Expertise at the beginning of the professional development comes in two forms. First, the professional development leaders provide expertise by planning a sequence of learning, facilitating activities, selecting classroom video for analysis, and guiding conversations among teachers. Second, classroom video featured during lesson analysis sessions features experienced teachers modeling the use of STeLLA lenses and strategies with their students. Although these experienced teachers are not present during the STeLLA sessions, they contribute their expertise to the participants by sharing their classroom practice, culture, and students with STeLLA participants.

In the STeLLA approach, guidance and support from experts slowly fade as teachers take on more and more of the responsibility for independently adopting and enacting the STeLLA strategies. After being introduced to the strategies and seeing them in action in video, teacher participants use a prepared set of STeLLA lessons that embed student thinking and Science Content Storyline strategies. Each teacher is filmed while teaching a portion of these lessons, and short clips are selected by a STeLLA facilitator for the small group to analyze. Participating teachers, at this point in the program, are very familiar with analyzing video from other teachers' classrooms and can now use the process to support and challenge one another and to negotiate deeper levels of understanding about the content, the strategies, and enactment in each participant's context.

In the final phase of the STeLLA approach, teachers experience the entire process of planning and enacting lessons that demonstrate their knowledge and use of the STeLLA lenses and strategies by collaboratively planning an instructional unit in a new science content area. STeLLA PD facilitators provide tools and processes that guide this planning and enactment, but step back from their role as leaders as the teachers step up to the challenge. The STeLLA program structure embodies the cognitive apprenticeship learning model by (1) working with experts and learning from the classroom video of more experienced STeLLA teachers, (2) trying out the STeLLA approach for themselves using prepared lesson plans and reflecting within a small group using video of their own initial attempts at using the strategies, and (3) planning new lessons with STeLLA peers with less input, guidance, or expertise from the PD leaders.

1.5.3   The Context of Teacher Learning: Video Analysis

The core teacher learning activity in all programs using the STeLLA approach is analysis of practice using grade-specific classroom video. Watching classroom video provides an opportunity for participants to see the same segment of classroom instruction and discuss the clip to negotiate a common understanding and language for communicating about teaching and learning. Videos provide a means of slowing down the action and focusing on specific aspects of the instruction for analysis. They can be watched more than once and focus on different elements of the teaching or student ideas being expressed. One of the design principles of the STeLLA approach is to provide written transcripts of each clip with time stamps so that teachers can cite evidence from the transcript to support claims they are making about the teaching strategies they see, the ideas students have about the science, or the particular times where teachers missed an opportunity to capitalize on student thinking to guide learning.

In line with situated cognition theory of teacher learning, it is essential that teachers analyze video featuring science content and classrooms at their grade level. This assures that each teacher is working in a meaningful, authentic context. The STeLLA leaders create video cases and select clips that will engage teachers in developing their initial understandings of (a) the science content, (b) the STeLLA teaching strategies, and (c) the video analysis process. Later in the program, teachers analyze video clips from their own and their colleagues' classrooms. Through these familiar contexts, teachers discuss their own students' unique ways of thinking about the content and ways to address the needs of their students in learning science content.

Using the STeLLA approach to teacher learning, program structure, and video analysis, teachers develop a level of PCK that links effective teaching practices to science content taught at their grade level. Participants analyze and reflect on their own use of the STeLLA strategies with their own students to develop a deeply contextualized understanding of the impact the STeLLA strategies can have on student learning. They have opportunities to try out the STeLLA lenses and strategies with a high degree of support that fades over time as they enact effective practices with greater independence and flexibility in their own classrooms.

1.6 Measuring PCK

1.6.1 The Need to Measure PCK

As PCK emerged as an important construct in science education, so did the need to measure it. As with the STeLLA studies described here, PCK is an intended outcome of many studies of professional development and teacher education interventions, and thus valid and reliable measures are needed. Such measures also have formative value during professional development, providing PD leaders with information on what participants do and do not know. Further, recent movements to develop national indicators in math and science education have recognized the importance of measuring teachers' knowledge of how to effectively teach their content (NRC 2013). That is, in order to monitor progress toward national recommendations for successful K-12 STEM education, we need to be able to measure growth in teachers' professional knowledge. Finally, measuring PCK is important because "assessments operationalize constructs" (William 2010). Instrument development work across science education has been foundational in developing frameworks that establish the structure of previously fuzzy concepts and making empirical observation possible. In doing so, the value of the construct to the field is increased, because if one cannot effectively measure something, it is unlikely to be valued or planned as an outcome of interventions.

1.6.2   Approaches to Measuring PCK

It has been said that as a field, we've been better at developing PCK than measuring it, which in part is due to a lack of consensus about the construct (Smith and Banilower 2015). While that is undoubtedly the case, several promising approaches have continued to be developed and refined, especially as the consensus model from the PCK Summit gains traction. Most measures situate their approach in the assertion that since PCK is the professional knowledge of teachers, it is visible in the professional work of teachers, which involves planning, teaching, and reflecting. *Planning measures* include instruments like the CoReS (Content Representations) and PaP-eRs (Pedagogical and Professional Experience Repertoires) developed by Loughran and colleagues (Loughran et al. 2012). These measures require teachers to write about the topics they are going to teach and what their lessons will involve and to describe reasons for those instructional decisions. *Teaching measures* use teachers' classroom practice as evidence of their PCK and use classroom observations or video to collect data. Finally, *reflecting measures* occur following instruction and involve teachers discussing the reasons for instructional decisions, sometimes with reference back to incidents in classroom video.

1.6.3   Measuring PCK Through Analysis of Practice

Our approach to measuring PCK diverges from this planning-teaching-reflecting cycle, in that a teacher's analysis of another teacher's classroom video is used as evidence of their PCK. While one could argue that such analysis falls outside of the regular professional work of teachers, and is therefore not professional knowledge one should expect a teacher to have, we believe it reasonable to expect that if a teacher has high PCK, they should be able to identify and describe effective science instruction and to identify student understanding from student talk. The instrument involves teachers watching a series of 5- to 10-min video clips, chosen to reveal different aspects of PCK. Teachers are asked to write about the teaching and learning visible in each video and provide analytic comments. Each response is scored using a rubric that codes for attention to the 16 STeLLA strategies, as well 8 distal codes that address more general aspects of PCK such as content accuracy, student thinking, and instructional coherence. Each code is scored 0 (no mention of the strategy),

1 (strategy is mentioned but not elaborated), or 2 (an elaborated discussion of the strategy is provided). This approach to measuring PCK emerged from the analysis-of-practice focus of the STeLLA model described above. We conclude this chapter by describing two studies that have used this instrument to measure the impact of teacher education and professional development on teacher PCK.

1.6.4 Two Studies Examining the Impact of Analysis of Practice on Teacher PCK

1.6.4.1 A Cluster Randomized Trial of In-Service Elementary Teachers

In this study, we tested the STeLLA professional development program in a cluster randomized trial with 140 fourth and fifth grade teacher participants in Colorado. Teachers were randomly assigned at the school level to either receive the STeLLA Lesson Analysis program or receive the same number of hours (88.5) of professional development focused entirely on deepening their content knowledge. This comparison was chosen because many studies and national reports argue that elementary teachers require this science content-focused PD to teach science more effectively. The structure of the STeLLA PD program is described above. We collected teacher responses to a science content assessment and the video analysis task before and after the program, as well as pretest and posttest student content assessments. The post-intervention science achievement for teachers and students was significantly higher in the STeLLA program than in the content deepening program (controlled for pre-intervention achievement). At the student level, the p value was 0.001, with a corresponding effect size of 0.68 (Hedges g) on a measure of students' science achievement (Taylor et al. 2016).

1.6.4.2 ViSTA Plus – A Quasi-experiment in Pre-service Teacher Education

The ViSTA Plus (Videocases for Science Teaching Analysis Plus) project was a 3-year study that followed a cohort of pre-service teachers from the science methods course, into student teaching, and finally through the first year of teaching. Methods courses from two universities in the Southwestern United States were assigned to either the ViSTA Plus treatment condition or a business-as-usual condition. In ViSTA Plus, we took the STeLLA framework, tools, and resources and adapted them for elementary pre-service teachers. The program includes a semester-long analysis-of-practice methods course that introduced pre-service teachers to STeLLA strategies and video-based lesson analysis and prepares them for participation in study groups that continued to meet synchronously online during student teaching and the first year of teaching.

The practice-focused science methods course is organized around the STeLLA conceptual framework and is intended to immerse pre-service teachers in learning in the context of the science classroom from the beginning of the program. First, pre-service teachers analyze videos and student work from experienced teachers' classrooms to support their learning about science and about the STeLLA Student Thinking and Science Content Storyline lenses and about the STeLLA teaching strategies. During student teaching, they teach model lesson plans embodying the STeLLA lenses and strategies and participate in lesson analysis study groups. The study group work continues during the first year of teaching as teams of teachers collaboratively design a series of grade-specific lessons and teach them in their own classrooms and again analyze and learn from one another's enactment of the lessons in synchronous online study groups. Similar to the STeLLA 2 study, we asked the ViSTA Plus teachers to teach a STeLLA-focused lesson sequence with the students in their student teaching classrooms and first year classrooms and comparison teachers to teach lessons with matched learning goals. Teacher content knowledge and PCK were assessed at four time points over the course of the study: pre-methods course, post-methods course, post-student teaching, and post-first year teaching. Due to significant attrition in the final stages of the project, only data from the first three time points is presented here. The students of the ViSTA Plus teachers were assessed on their science content knowledge before and after each teaching unit.

1.7   Findings: Teacher PCK

As we described above, teacher PCK was measured using a video analysis task measure in the STeLLA2 and ViSTA Plus studies. Growth on this outcome was observed for teachers in both treatment and comparison conditions over the course of the interventions. STeLLA and ViSTA Plus teachers outperformed the comparison condition teachers in both studies (Fig. 1.2a and b). Of particular interest, in the

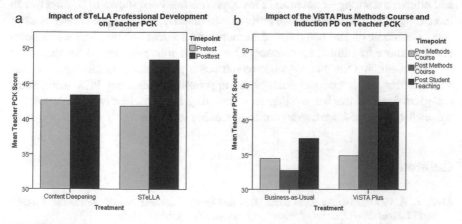

Fig. 1.2 (**a**) and (**b**) The impact of lesson analysis professional development on teacher PCK in the STeLLA2 and ViSTA Plus studies

ViSTA Plus study of pre-service teachers, we observed the ViSTA Plus teachers increased their PCK significantly during their methods course, whereas teachers in the traditional methods course did not make significant gains in PCK until their student teaching experience. This finding speaks to the value of bringing classroom video and lesson analysis into the methods course, situating learning in the classroom context, and the development of PCK requiring professional experiences.

1.8 Findings: Connecting PCK to Student Content Knowledge

In both studies, the treatment condition significantly impacted student content knowledge (Taylor et al. 2016; Wilson et al. 2017), which led to the question – to what extent was teacher PCK predictive of student learning? In both studies, PCK was a positive but not a significant mediator in models predicting student learning, after the impact of the treatment had been accounted for. The problem here might be one of power, in that both studies were powered to find differences in treatments with respect to teacher and student outcomes, not the relationship between the outcomes. We will continue to explore this relationship, since we believe PCK to be an important factor in providing students with effective learning experiences.

1.9 Conclusion and Future Directions

The theme of this book is the translation of PCK research into practice. In this chapter, we described an approach to professional development grounded in a model of PCK that centers around a teacher's knowledge of how to elicit and respond to student thinking, to engage students in the practices of science, and to develop coherent and effective learning experiences. This approach has been shown to be effective in increasing PCK in both in-service professional development and pre-service teacher education contexts and has resulted in increased student achievement when compared to more traditional approaches. Current and future directions for this work include examining the pathways between teacher professional knowledge and student learning, further development of our approach to measuring PCK using automated analysis of teacher writing, and exploring how to effectively scale up this somewhat resource-intensive approach to teacher learning.

References

Abell, S. A. (2008). Twenty years later: Does pedagogical content knowledge remain a useful idea? *International Journal of Science Education, 30*(10), 1405–1416.
Bertram, A. (2014). CoRes and PaP-eRs as a strategy for helping beginning primary teacher develop their pedagogical content knowledge. *Educación Química, 25*(3), 292–304.

Borko, H., & Livingston, C. (1989). Cognition and improvisation: Differences in mathematics instruction by expert and novice teachers. *American Educational Research Journal, 26*(4), 473–498.

Carlson, J., Stokes, L., Helms, J., Gess-Newsome, J., & Gardner, A. (2015). The PCK Summit: A process and structure for challenging current ideas, provoking future work, and considering new directions. In A. Berry, P. Friedrichsen, & J. Loughran (Eds.), *Re-examining pedagogical content knowledge in science education* (pp. 14–27). New York, NY: Routledge.

Cochran, K., DeRuiter, J., & King, R. (1993). Pedagogical content knowing: An integrative model for teacher preparation. *Journal of Teacher Education, 44*(4), 263–272.

Collins, A. (2006). Cognitive apprenticeship. In R. Keith Sawyer (Ed.), *The Cambridge handbook of the learning sciences*. Cambridge: Cambridge University Press.

Daehler, K. R., Heller, J. I., & Wang, N. (2015). Supporting growth of pedagogical content knowledge in science. In A. Berry, P. Friedrichsen, & J. Loughran (Eds.), *Re-examining pedagogical content knowledge in science education* (pp. 45–59). New York, NY: Routledge.

Friedrichesen, P., Lannin, J., & Sickel, A. (2012). *PCK development among science teachers*. Retrieved from http://pcksummit.bscs.org/sites/default/files/Friedrichsen_EP.pdf.

Gess-Newsome, J. (1999). Secondary teachers' knowledge and beliefs about subject matter and its impact on instruction. In J. Gess-Newsome & N. G. Lederman (Eds.), *Examining pedagogical content knowledge: The construct and its implications for science education* (pp. 51–94). Dordrecht, The Netherlands: Kluwer Academic Publishers.

Gess-Newsome, J. (2015). A model of teacher professional knowledge and skill including PCK. In A. Berry, P. Friedrichsen, & J. Loughran (Eds.), *Re-examining pedagogical content knowledge in science education* (pp. 28–42). New York, NY: Routledge.

Gess-Newsome, J., Taylor, J. A., Carlson, J., Gardner, A. L., Wilson, C. D., & Stuhlsatz, M. A. M. (2017). Teacher pedagogical content knowledge, practice, and student achievement. *International Journal of Science Education*, 1–20. Published online at https://doi.org/10.1080/09500693.2016.1265158.

Grossman, P. L. (1990). *The making of a teacher: Teacher knowledge and teacher education*. New York, NY: Teachers College Press.

Hashweh, M. Z. (1987). Effects of subject matter knowledge in the teaching of biology and physics. *Teaching and Teacher Education, 3*(2), 109–120.

Heller, J. I., Daehler, K. R., & Kaskowitz, S. R. (2004, April). *Fostering pedagogical content knowledge about electric circuits through case-based professional development*. Paper presented at the annual meeting of the National Association for Research in Science Teaching, Vancouver.

Hume, A. (2010, May). *CoRes as tools for promoting pedagogical content knowledge of novice science teachers*. Chemistry Education in New Zealand, 13–19.

Kind, V. (2009). Pedagogical content knowledge in science education: Perspectives and potential for progress. *Studies in Science Education, 45*(2), 169–204.

Kirschner, S., Borowski, A., & Fischer, H. (2012). *Development of a valid and reliable test to assess physics teachers' professional knowledge*. Retrieved from http://pcksummit.bscs.org/sites/default/files/Fischer%20EP.pdf.

Lave, J., & Wenger, E. (1999). Learning and pedagogy in communities of practice. In J. Leach & B. Moon (Eds.), *Learners and pedagogy* (pp. 21–33). London: Paul Chapman.

Loughran, J., Brown, J., & Doecke, B. (2001). Continuities and discontinuities: The transition from pre-service to first-year teaching. *Teachers and Teaching, 7*(1), 7–23.

Loughran, J., Berry, A., & Mulhall, P. (2012). Portraying PCK. In *Understanding and developing science teachers' pedagogical content knowledge* (pp. 15–23). Rotterdam: Sense Publishers.

Magnusson, S., Krajcik, J., & Borko, H. (1999). Nature, sources, and development of pedagogical content knowledge for science teaching. In J. Gess-Newsome & N. G. Lederman (Eds.), *Examining pedagogical content knowledge: The construct and its implications for science education* (pp. 95–132). Dordrecht, The Netherlands: Kluwer Academic Publishers.

National Research Council. (2013). *Monitoring progress toward successful K-12 STEM education: A nation advancing?* Committee on the Evaluation Framework for Successful K-12 STEM Education. Board on Science Education and Board on Testing and Assessment,

Division of Behavioral and Social Sciences and Education. Washington, DC: The National Academies Press.

Padilla, K., Ponce-de-León, A. M., Rembado, F. M., & Garritz, A. (2008). Undergraduate professors' pedagogical content knowledge: The case of 'amount of substance. *International Journal of Science Education, 30*(10), 1389–1404.

Park, S., & Oliver, J. S. (2008). Revisiting the conceptualisation of pedagogical content knowledge (PCK): PCK as a conceptual tool to understand teachers as professionals. *Research in Science Education, 38*(3), 261–284.

Rollnick, M., & Mavhunga, E. (2012). *Mixed-methods investigation of physical science teachers' PCK*. Retrieved from http://pcksummit.bscs.org/sites/default/files/Rollnick.EP__0.pdf.

Rollnick, M., Bennett, J., Rhemtula, M., Dharsey, N., & Ndlovu, T. (2008). The place of subject matter knowledge in pedagogical content knowledge: A case study of South African teachers teaching the amount of substance and chemical equilibrium. *International Journal of Science Education, 30*(10), 1365–1387.

Roth, K. J., Garnier, H., Chen, C., Lemmens, M., Schwille, K., & Wickler, N. I. Z. (2011). Videobased lesson analysis: Effective science PD for teacher and student learning. *Journal of Research in Science Teaching, 48*(2), 117–148.

Schneider, R. M., & Plasman, K. (2011). Science teacher learning progressions: A review of science teachers' pedagogical content knowledge development. *Review of Educational Research, 81*(4), 530–565.

Shulman, L. S. (1987). Knowledge and teaching: Foundations of the new reform. *Harvard Educational Review, 57*(1), 1–22.

Shulman, L. S. (2002). Truth and consequences? Inquiry and policy in research on teacher education. *Journal of Teacher Education, 53*(3), 248–253.

Shulman, L. S. (2015). PCK: Its genesis and exodus. In A. Berry, P. Friedrichsen, & J. Loughran (Eds.), *Re-examining pedagogical content knowledge in science education* (pp. 28–42). New York, NY: Routledge.

Smith, S., & Banilower, E. (2012). *Challenges of developing PCK measures for large-scale studies*. Retrieved from http://pcksummit.bscs.org/sites/default/files/Smith%20Banilower%20EP.pdf.

Smith, P. S., & Banilower, E. R. (2015). A new application of the uncertainty principle. In A. Berry, P. Friedrichsen, & J. Loughran (Eds.), *Re-examining pedagogical content knowledge in science education* (pp. 28–42). New York, NY: Routledge.

Taylor, J. A., Roth, K., Wilson, C. D., Stuhlsatz, M. A., & Tipton, E. (2016). The effect of an analysis-of-practice, videocase-based, teacher professional development program on elementary students' science achievement. *Journal of Research on Educational Effectiveness, 10*(2), 241–271.

van Driel, J., & Henze, I. (2012). *Development of PCK among pre-service and in-service science teachers*. Retrieved from http://pcksummit.bscs.org/sites/default/files/VanDriel.EP_.pdf.

Van Driel, J. H., Verloop, N., & de Vos, W. (1998). Developing science teachers' pedagogical content knowledge. *Journal of Research in Science Teaching, 35*(6), 673–695.

William, D. (2010). What counts as evidence of educational achievement? The role of constructs in the pursuit of equity in assessment. *Review of Research in Education, 34*, 254–284.

Williams, J. (2012). Using CoRes to develop the pedagogical content knowledge (PCK) of early career science and technology teachers. *Journal of Technology Education, 24*(1), 34–53.

Wilson, C., Stuhlsatz, M., Hvidsten, C., & Stennett, B. (2017, April). *Examining the impact of lesson-analysis based teacher education across methods courses, student teaching, and induction*. Paper presented at the annual meeting of NARST, San Antonio, TX.

Chapter 2
The Intertwined Roles of Teacher Content Knowledge and Knowledge of Scientific Practices in Support of a Science Learning Community

Lane Seeley, Eugenia Etkina, and Stamatis Vokos

Abstract In this chapter we envision the science classroom as an authentic scientific community. In this vision, student ideas can influence the trajectory of scientific investigation. Teachers serve as experts and guides, but they also can learn alongside their students. To do this, they need to listen to the students and be able to build on students' original ideas to help them learn. What knowledge does a teacher draw on in such a classroom? In this chapter we empirically investigate some ways in which a teacher can utilize both knowledge of the subject matter and knowledge of science practices to respond productively to student thinking. We present data from a large study of knowledge for teaching energy. The subjects of this study were high school physics teachers. We found that in some instructional situations, teachers with insufficient content knowledge cannot productively respond to student reasoning. We also found cases where teachers can compensate for lack of content knowledge if they are skilled in science practices. To explain our findings, we hypothesize the existence of two types of content knowledge: foundational content knowledge and elaborative content knowledge. Furthermore, we suggest that foundational content knowledge along with knowledge of scientific practices can allow teachers to compensate for insufficient elaborative content knowledge. We discuss the implications of our hypothesis for future research and for the preparation and professional development of physics teachers.

L. Seeley (✉)
Department of Physics, Seattle Pacific University, Seattle, WA, USA

E. Etkina
Department of Learning and Teaching, Rutgers University Graduate School of Education, New Brunswick, NJ, USA

S. Vokos
Physics Department, California Polytechnic State University, San Luis Obispo, CA, USA

© Springer International Publishing AG, part of Springer Nature 2018
S. M. Uzzo et al. (eds.), *Pedagogical Content Knowledge in STEM*,
Advances in STEM Education, https://doi.org/10.1007/978-3-319-97475-0_2

Keywords PCK · High school physics · Foundational content knowledge · Elaborative content knowledge · Pre-service learning · Teacher professional development · Content knowledge for teaching · Content knowledge for teaching energy · Disciplinary content knowledge · Systems

2.1   Introduction

Ms. Cordova is preparing for her high school physics class tomorrow. She has noticed that her students have used the terms for various forms of energy, but the evidence she has collected suggests that so far, only kinetic energy and gravitational potential energy are being associated reliably with observable indicators. As she was leaving class today, a student group mentioned to Ms. Cordova that during a process in which a kicked ball rolls to a stop, its kinetic energy gradually decreases because its speed decreases, and therefore it must be transforming into potential energy. "It's potential kinetic energy," one of the students in the group exclaimed as she went out the door.

Ms. Cordova ponders various instructional moves to respond productively to the students' ideas. She notes the students' facility with connecting kinetic energy to its indicator—the speed of the ball. She appreciates the group's intellectual commitment to the idea that energy conservation requires that the decrease in a form of energy in a system be accounted. She is realizing that the imperceptibility of temperature changes in interacting surfaces in many physical processes does not lend itself to students' thinking about thermal energy as a likely increasing form of energy that compensates for a form of energy that is decreasing perceptibly. She is concerned that if the class does not recognize the role of thermal energy in lots of physical phenomena, her students might not be able to connect their school learning of "conservation of energy" to sociopolitical issues associated with efforts to "conserve energy" (Daane et al. 2014) or even to energy learning in other science disciplines.

The preceding vignette is inspired by multiple classroom discussions in which learners spontaneously bring up the phrase, "potential kinetic energy." To respond productively to student ideas, Ms. Cordova needs to marshal knowledge that is strongly dependent on the specific topic her students are learning. In this example, she needs to know enough (and be curious about her students' ideas) to notice the disciplinary substance of the offhand comment her students made. She needs to know that the energy of the ball, the floor, and the air will remain constant as no energy is flowing into or out of the system of the three objects and that the interaction between the ball and the ground converts the kinetic energy of the ball into thermal energy of the ball, the floor, and the air. She also needs to know that in many mechanical contexts, temperature changes are too small to have been experienced by her students in their daily lives. She needs to know of mesoscopic and microscopic models for friction (Besson and Viennot 2004) that can help her students develop a causal picture for the "heating up" associated with surfaces rubbing against each other; she also needs to be aware that learners tend to conflate models of processes at different scales (microscopic/macroscopic), whereas even seasoned energy learners have a hard time separating cleanly effects that involve both ordered macroscopic motion (e.g., the "wind" produced by the ball moving through the air) and disordered microscopic motion (e.g., during the dying-down of air currents).

The teacher needs to know and prioritize sequences of phenomena whose energy analyses build on the previous ones. She needs to know ways in which mathematization can serve as a tool rather than an impediment to conceptual development in this context. She needs to know multiple energy tracking representations to help her students gain more insight. She needs to know of experiments that are possible to conduct in the classroom and the affordances and limitations of each for specific instructional purposes. In the case that students do not spontaneously bring up the ideas described above, the teacher needs to know enough to judge whether or not eliciting such ideas through a rich question that she will pose is likely to serve her immediate and ultimate learning goals for the students.

The preceding episode illustrates the complexity of the knowledge that is in play when a teacher is committed to responding productively to student ideas.[1] It is not enough for a teacher to "know the content" and "know how to teach." A successful teacher must negotiate pedagogical decisions that are inextricably linked to the content her students are learning. L. Schulman introduced the construct of pedagogical content knowledge (PCK) in order to acknowledge and study the fundamental interaction and interdependence of pedagogical and content knowledge (Shulman 1986, 1987). During the past three decades, there has been significant progress in defining, categorizing, and assessing PCK domains (Berry et al. 2015; Gess-Newsome 2011).

There are several instruments that assess teacher's PCK in specific physics domains (Keller et al. 2017). The Magnusson, Krajcik, and Borko model of science PCK (Magnusson et al. 1999) was the first attempt to detail this knowledge, and recently a new, revised model of PCK has emerged (Gess-Newsome 2015). In this model PCK is just one component of teacher professional knowledge and practice. Most importantly, in this model teacher professional knowledge is not just subject-specific; it is topic-specific.

In this study, we draw on a somewhat different model of teacher knowledge—content knowledge for teaching (CKT) (Ball et al. 2008). CKT is operationalized as the specialized disciplinary knowledge that is specifically relevant to the work of teaching. CKT is broader than PCK because it includes both disciplinary knowledge and pedagogical applications of that knowledge. CKT also has a strong empirical focus on the specialized disciplinary knowledge that teachers actually do draw on in real classrooms. In this study we adapt and apply this model (i.e., develop a theoretical framework) to one narrow domain of high school physics and show how to use this theoretical framework to assess teacher knowledge for teaching energy in a first physics course in the context of mechanics. Specifically, in this chapter we aim to answer the following research questions:

- To what degree is the productivity of teacher responses to student thinking contingent on the teacher's disciplinary content knowledge of the relevant physics topic?

---

[1] We do not intend to provide an exhaustive description of the knowledge needed to respond productively to student ideas. In particular, additional knowledge and skills and dispositions are surely needed to engage productively moment by moment with real students in a real classroom. We concentrate on important aspects of the disciplinary knowledge that is deployed by the teacher in service of energy instruction.

- Are there specific aspects of disciplinary content knowledge of the relevant topic that are critical for supporting student thinking?

In order to address these questions empirically, we will first briefly introduce CKT as a framework of assessing teacher knowledge. Next, we will describe a written assessment of CKT in the narrow domain of high school physics energy instruction. Finally, we will analyze general and specific patterns of teacher responses on this assessment that provide insight into the preceding questions.

2.2 Content Knowledge for Teaching (CKT)

2.2.1 Tasks of Teaching and Student Energy Targets

The work described in this chapter is one of the products of a multi-year effort to develop and validate a set of substantively coherent measures that assess CKT in physics in the domain of energy, through both tests and evidence from instructional practice. The project focused on one conceptual area so as to forge a tight theoretical and empirical link between CKT and practice.

To establish this link, we developed the domain model through an extensive review of the literature and through observations of expert teachers teaching energy. The domain model of content knowledge for teaching energy (CKT-E) involves two components. The first component is the critical *tasks of teaching* (ToTs) (Ball 2000). Tasks of teaching describe the key activities through which teachers and students enact practices that promote and support student learning. For our project we developed the following list of broad categories of the tasks of teaching (each has several subcategories listed in Appendix A): (I) anticipating student thinking around science ideas; (II) designing, selecting, and sequencing learning experiences and activities; (III) monitoring, interpreting, and acting on student thinking; (IV) scaffolding meaningful engagement in a science learning community; (V) explaining and using examples, models, representations, and arguments to support students' scientific understanding; and (VI) using experiments to construct, test, and apply concepts.

Although we do not expect that teachers engage in all tasks of teaching in every lesson, we should be able to observe a teacher engaged in each of those tasks many times during teaching of the energy unit. Further, while these tasks of teaching are not the only tasks in which teachers engage while teaching, the CKT theory assumes that for students to learn, teachers should engage in all of these tasks across each unit of instruction (Ball 2000; Gitomer et al. 2014).

The second component of our domain model is the *student energy targets* (SETs). It focuses on the specific content targets of energy in mechanics contexts for students and articulates features that are important in the domain (in our case, energy taught in the context of mechanics in a typical high school physics course), including concepts and skills, critical tasks in which those are manifest, and knowledge representations. Targets are separated into several broad categories: (A) connections of energy and everyday experiences (National Research Council 1999, 2000, 2012), (B) choice of system (Van Heuvelen and Zou 2001; Lindsey 2014; Lindsey et al. 2009, 2012), (C) identification of and differentiation between energy and other physics concepts (Swackhamer 2005; Brewe 2011), (D) transfer of energy (Daane et al. 2013, 2014, 2015; Scherr et al. 2012a, b, 2013a, b; Seeley et al. 2014), (E) use of mathematics, (F) use of representations (Daane et al. 2013, 2014, 2015; Van Heuvelen and Zou 2001; Scherr et al. 2012a, b, 2013a, b), and (G) use of science practices (Etkina 2015). Elaborations of each of these categories are provided in Appendix B.

2.2.2   CKT-E Residing at the Intersection of ToTs with SETs

We conceptualize CKT-E as "residing" at the intersection of specific tasks of teaching with the content learning targets—in our case, student energy targets. In essence, we ask what knowledge a teacher would need to "have" to execute a particular task of teaching in the domain of energy to support a particular student energy target. We also recognize that when teachers are able to respond to the scientific ideas and questions of their students, their responses will often recruit knowledge that lies beyond the learning targets they have set for their students. In these cases teachers may draw on horizon knowledge (Ball 2000). The example of Ms. Cordova illustrates how any instructional situation that results from specific ToTs in support of specific SETs can reveal a broad array of CKT.

Ms. Cordova has specific student energy targets in mind. She wants her students to *recognize the important role of internal energy in interpreting or explaining everyday phenomena* (SET A2). She also wants to help them *understand that energy cannot be observed directly and know[s] how different forms of energy correspond to different measurable physical quantities* (SET A1). Finally, she wants them to realize that *equal amounts of energy in different forms are not equally perceptible* (horizon knowledge). The clear and visible motion energy of a rolling ball is much more perceptible than an equal quantity of thermal energy in the ball/environment when the ball rolls to a stop.

In service of these important student energy targets, Ms. Cordova undertakes several different tasks of teaching. She is attentive to what her students say about "potential kinetic energy" because she was already *anticipating specific student challenges related to constructing scientific concepts* (ToT Ia) and *anticipating likely partial conceptions and alternate conceptions* (ToT Ib). She needs to understand the ideas behind her student's words and *interpret both productive and problematic aspects of her students thinking* (ToT IIIb). Ms. Cordova anticipates and interprets in order to inform her actions. She is also thinking of what to do the next day to help students "see" the conversion of kinetic energy into internal thermal energy. She wants to *design, select, and sequence a learning experience that will address her students' actual learning trajectories by building on productive elements and addressing problematic ones* (ToT IId).

Ms. Cordova could design a number of different learning experiences to help her students move toward the energy learning targets she has set out for them. For example, she could use an infrared camera or an infrared video to reveal evidence of thermal energy (Vollmer and Mollmann 2010) and use this evidence to inform their discussion. Alternatively, she could explore her student's suggestion of "potential kinetic energy" and help the student unpack the ideas behind that phrase. A productive instructional response will draw on domain-specific knowledge. We operationalize CKT as the knowledge that a teacher is likely to draw on in her efforts to carry out specific ToTs in service of specific SETs. Operationalized in this way, CKT extends beyond the student learning targets and includes both disciplinary knowledge and pedagogical knowledge. For example, disciplinary knowledge of infrared photography belongs to what Ball calls "horizon knowledge." Horizon knowledge is beyond the scope of student energy targets but still relevant for a productive instructional response. On the other hand, anticipating why "potential kinetic energy" might make sense to students in this scenario is a pedagogical challenge that is specific to this disciplinary context.

2.3 Assessing CKT

The specific CKT a teacher recruits will depend on the SETs, the ToTs, the instructional situation, and the instructional decisions. Therefore, fully listing all examples of CKT, even in a narrow domain, is not productive. In constructing a written CKT assessment, we have focused on a representative subset of ToT/SET combinations as instantiated through various instructional scenarios. Some of the items assess disciplinary knowledge of physics, which may be relevant for a teaching situation but does not require detailed knowledge of student learning or of the school context to be answered correctly. We designate these items as content knowledge for teaching-disciplinary (CKT-D). In addition to assessing disciplinary knowledge, some items require an understanding of content-specific learning trajectories and pedagogical strategies. We designate these items as content knowledge for teaching-pedagogical (CKT-P). We do not see these distinctions as an effort to measure distinct domains of knowledge. Rather, our goal was to ensure that the tasks we developed represented a range of disciplinary and pedagogical challenges.

The details of test construction, piloting, and revisions are provided elsewhere (Etkina et al. 2018). Here we only describe the final test. The final form of the assessment contained 26 scored items that were associated with 15 unique teaching scenarios. The form included 6 constructed-response (CR) items scored on 3- or 4-point scales, 6 polytomous items in which test-takers could correctly answer between 0 and 5 or 6 items, and 14 multiple-choice items that were dichotomously scored. Every item contained a rationale that was directly connected with specific tasks of teaching and the student energy targets.

2.3.1   Example Items for Assessing Disciplinary Content Knowledge for Teaching (CKT-D) and Pedagogical Content Knowledge for Teaching (CKT-P)

Examples of CKT-D and CKT-P items are provided in Figs. 2.1 and 2.2. The item shown in Fig. 2.1 is labeled as *Cyclist, CKT-D, SR. Cyclist* describes the context of the item. *CKT-D* identifies this item as primarily assessing disciplinary knowledge of physics. In other words, correctly answering this item requires no knowledge of

In a situation with a number of interacting objects, one may select any subset of them as the system of interest. The objects that have not been selected as belonging to the chosen system are therefore external to the system.

Ms. Inez wants to help her students recognize that energy is a conserved quantity but that the energy of a particular system may not be constant, depending on the specific scenario and the choice of system for analysis. She decides to have them focus on the scenario of a cyclist riding up a hill at constant speed.

For each of the following, systems indicate whether the energy associated with that system increases, decreases, or remains approximately constant.

	Increases	Decreases	Approximately Constant
A. Bicycle, rider, air, pavement and Earth			
B. Bicycle, rider and Earth			
C. Bicycle, air and pavement			
D. Bicycle and Earth			

Fig. 2.1 *Cyclist, CKT-D, SR* item

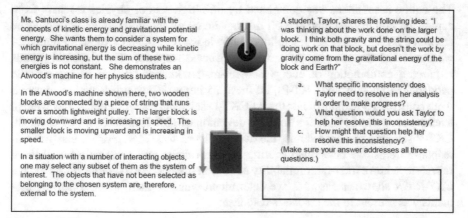

Ms. Santucci's class is already familiar with the concepts of kinetic energy and gravitational potential energy. She wants them to consider a system for which gravitational energy is decreasing while kinetic energy is increasing, but the sum of these two energies is not constant. She demonstrates an Atwood's machine for her physics students.

In the Atwood's machine shown here, two wooden blocks are connected by a piece of string that runs over a smooth lightweight pulley. The larger block is moving downward and is increasing in speed. The smaller block is moving upward and is increasing in speed.

In a situation with a number of interacting objects, one may select any subset of them as the system of interest. The objects that have not been selected as belonging to the chosen system are, therefore, external to the system.

A student, Taylor, shares the following idea: "I was thinking about the work done on the larger block. I think both gravity and the string could be doing work on that block, but doesn't the work by gravity come from the gravitational energy of the block and Earth?"

a. What specific inconsistency does Taylor need to resolve in her analysis in order to make progress?
b. What question would you ask Taylor to help her resolve this inconsistency?
c. How might that question help her resolve this inconsistency?
(Make sure your answer addresses all three questions.)

Fig. 2.2 *Atwood's, CKT-P, CR* item

students or pedagogy. *SR* indicates that this item is a selected-response or multiple-choice item rather than a constructed-response item. While this item is not pedagogical in nature, it does assess an area of disciplinary knowledge that is particularly relevant to the work of teachers. In order for teachers to give their students full ownership of the energy analysis process, they will need to allow their students to select the system for energy analysis. When teachers do this, they must be prepared to recognize how different student-generated system choices will affect the energy analysis. In Sect. 2.4 we discuss the important role of systems in energy analysis.

Atwood's, CKT-P, CR, shown in Fig. 2.2, is an item that presents the teacher with a pedagogical challenge and recruits a constructed response. This item is expected to require sophisticated disciplinary knowledge, but it cannot be successfully answered based on disciplinary knowledge alone. The teacher must use her disciplinary knowledge to interpret and evaluate a student statement, plan her instructional response, and anticipate how the student will respond.

While the preceding two items can be readily categorized, not all of the items on this assessment are as explicitly CKT-D or CKT-P. Rather, the assessment items are distributed along a continuum of increasing pedagogical challenge. The items that we have categorized as CKT-P are those for which the principal cognitive challenge is directly related to pedagogy. We should also clarify that items that we have classified as CKT-P should not be considered to involve less sophisticated physics knowledge. Many of the CKT-P items require physics knowledge, both sophisticated and subtle.

2.3.2 General Patterns in Teachers' Performance on CKT-D and CKT-P Items

The online assessment was completed by 362 high school physics teachers from across the country. Among the 50 distinct items on the assessment, 25 could be answered based largely on disciplinary knowledge (CKT-D). Of these 25 items, 24 were selected-response items; 1 was constructed response. The remaining items required a combination of energy subject-matter knowledge and energy-specific pedagogical knowledge (CKT-P). Of these 25 items, 20 were selected response; 5 were constructed response. For the 24 CKT-D selected-response items, the average teacher score was 64% with a standard deviation of 18% (see Table 2.1). For the 20 CKT-P selected-response items, the average score was 71%. For all constructed-response items, we developed scoring 3-point rubrics and iteratively found these rubrics to achieve inter-rater reliability of 90% or greater. For example, on *Atwood's, CKT-P, CR* shown in Fig. 2.2, we determined whether each of the following elements was present in the teacher's response:

- Personally responsive – Teacher poses a question which is responsive to Taylor's question in a way that might lead her to move forward productively with her thinking.

Table 2.1 Participant performance on test items by category, $N = 362$

Category	# Items (points possible)	Percentage of points possible
CKT-D, SR	24 (24)	64% ± 18%
CKT-P, SR	20 (20)	71% ± 13%
CKT-D, CR	1 (3)	57% ± 38%
CKT-P, CR	5 (15)	49% ± 20%

- Intellectually responsive – Teacher poses a question that is responsive to Taylor's intellectual need to precisely define her system when analyzing the work done by the gravitational force and changes in gravitational potential energy.
- Productive – Teacher poses a question that can be answered and is likely to lead to increased understanding.

Responses were awarded a single point for each element identified for a maximum possible score of 3. On the single CKT-D constructed-response item, the average score was 57%. For the 5 CKT-P constructed-response items, the average score was 49%. Standard deviations of teacher scores are also provided in Table 2.1. On the basis of these findings, we see that the teachers found both CKT-D and CKT-P items reasonably challenging and that the teachers achieved similar success rates.

2.3.3 Contingency of CKT-P on CKT-D Among Teachers and Non-teachers

The CKT-P items on this assessment involve the application of disciplinary knowledge to address a pedagogical challenge. Therefore, it is reasonable to assume that some level of CKT-D would be required for teachers to perform well on the CKT-P items. We also expect that teachers would demonstrate a higher level of CKT-P compared with non-teachers with similar overall disciplinary knowledge of physics. To test the latter expectation, we have previously reported the use of a cognitive diagnostic model (Phelps et al. in preparation). CDMs are a type of confirmatory latent class model. Fitting a CDM to the data requires specifying which skills are evaluated by each item, which in turn allows the specification of the latent classes. Here we hypothesize that examinees belong to one of three latent classes—(1) neither type of CKT, (2) CKT-D but not CKT-P, and (3) both CKT-D and CKT-P. These latent classes were constrained to be ordered, which reflected our understanding of the hierarchical nature of CKT. That is, for an examinee to have mastered the CKT-P skill, they would also have to have mastered CKT-D. After the specification of the latent classes and their structure, the CDM was fit to the data. CDMs, and latent class models more generally, use the pattern of item responses to assign a probability of each examinee belonging to each latent class. These latent class assignments maximize the likelihood of the observed item responses in a process known as maximum likelihood estimation. The latent class categorizations are summarized in

Table 2.2 Categorization of test subjects according to CKT-D and CKT-P

CKT disciplinary	CKT pedagogy		
		Low = 0	High = 1
	Low = 0	00	
	High = 1	10	11

Table 2.3 Latent class categorization of teachers and non-teachers, N teachers = 362; N non-teachers = 311

	Latent class 00	Latent class 10	Latent class 11
Both	0.59	0.14	0.27
Teachers (N = 362)	0.44	0.05	0.51
Non-teachers (N = 311)	0.75	0.25	0.00

Table 2.2. This CDM was then used to compare the population of physics teachers with a population of 311 physics majors. The results are shown in Table 2.3.

Among the teacher sample group, 51% showed mastery of both CKT-D and CKT-P. In contrast, there were no subjects within the non-teacher group who demonstrated mastery of CKT-P. While the teacher group performed better on the overall assessment, this difference alone does not explain the difference in their mastery of CKT-P. These results strongly suggest that teachers develop CKT-P through their work as teachers. The results also suggest a complex relationship between disciplinary knowledge and the application of that disciplinary knowledge in service of a pedagogical challenge. In the next two sections, we will explore that relationship in greater detail. We will specifically focus on the role of supporting disciplinary knowledge as a resource for productively attending to student reasoning.

2.4 Patterns in Teacher System Energy Reasoning

The example items previously shown in Figs. 2.1 and 2.2 are similar in that they both involve systems reasoning. Systems reasoning in physics involves the ability to strategically select a system for analysis and recognize that the system choice will determine if the energy of that system is constant for a given scenario, recognizing that if the system was defined differently, its energy might not have been constant (Lindsey 2014; Lindsey et al. 2009, 2012). On the uphill cycling items, many of the teachers taking the test had difficulty interpreting the different system choice options and how the system choice would affect the energy analysis, as suggested here:

> Generally the earth is not needed for inclusion into the calculation of the change in potential energy of the bike. Assuming that $U = mgh$ is being used, the mass of the earth and its subsequent motion is typically ignored. We only really want to analyze the forces and changes in energy of the rider/bike system. If we wanted to treat the Earth as the pavement, and use this in formulating our calculation of the work done by friction, then I suppose this

could be useful. This leaves out the air unfortunately, which contributes significantly as negative work on the biker, and doesn't help us answer questions about the system effectively.

This teacher does not appear to understand that using the equation, $U = mgh$, implicitly includes earth in the system for analysis. Several teachers explicitly commented that they did not understand what was meant by the word *system* and how to apply it. Based on those comments, the items were revised for clarity and included the following statement explicitly describing how we wanted teachers to interpret the word in responding to the assessment items: "In a situation with a number of interacting objects, one may select any subset of them as the system of interest. The objects that have not been selected as belonging to the chosen system are therefore external to the system."

Nevertheless, the role of a system remained difficult for many teachers who took our field test. We suspect that this difficulty can partly be attributed to disciplinary differences in the meaning of the word *system*. In physics, when analyzing a process, specifying a system is the prerogative of the scientist. Specifying a system involves deciding which object or objects to include in the system and which objects will be external to it. A choice of system in physics says nothing about the presence or absence of interactions among objects. That is additional information. In biology or ecology, a system includes all of the relevant objects together with their interactions. If you specify that an object is not within your system, in biology that is taken to mean that the object does not play a significant role in the processes under analysis. In physics, however, it just means that the object in question is in the environment of your chosen system. To get a better understanding of this distinction, compare the following:

- Consider the ecosystem of Yellowstone National Park. Now imagine that wolves are not part of your system. For the ecologist this would suggest that either the wolves have been removed from the park or they do not play a significant role within the ecosystem of the park.
- Now consider a person pushing a box on a rough floor. A physicist might strategically decide to include just the box and the floor in their system. In this case, the kinetic energy that is converted into thermal energy through the friction interaction would remain in the system. The person would still play a critical role in the energy story, but they would be transferring energy to the system through the process of work. By assigning the person to the environment, the physicist is choosing not to track the complex changes in chemical and thermal energy within the person.

The canonical approach to energy analysis in physics is contingent on using the second approach to systems. When the concept of work is introduced to quantify energy transfer into or out of a chosen system, this idea implicitly recognizes that objects in the environment are having a significant influence on the chosen system.

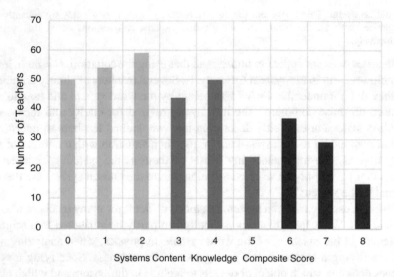

Fig. 2.3 Teachers' systems reasoning composite CKT-D scores

2.4.1 Teacher Performance on Systems Reasoning Items

Systems reasoning is a foundational aspect of energy reasoning. Therefore, we designed several assessment items that directly assess a teacher's ability to apply system-based disciplinary energy reasoning in an instructional context. The uphill cycling item shown in Fig. 2.1 provides an example of one of these items in which teachers are asked to match an energy description with the corresponding system choice. From these items we created an 8-point composite index of CKT-D for energy systems reasoning and used it to score teachers' work. Teachers' scores on this composite index varied widely as shown in Fig. 2.3. We classified 163 teachers who scored 3 or less on this composite index as demonstrating low systems CKT-D and 81 teachers who scored 6 or higher as demonstrating high systems CKT-D. The distribution of teacher scores on this composite index illustrates that systems is an area of CKT-D where teachers exhibit wide disparities in understanding.

2.4.2 Attending Productively to Systems Reasoning

Teaching responsively involves sifting through the multitude of ideas that students voice and recognizing those ideas that provide entry points for additional scientific reasoning (Robertson et al. 2016). Several items on this assessment present hypothetical classroom situations in which students have expressed scientific ideas that are incomplete yet potentially productive. We then ask teachers to select an instructional response from a list of options and/or describe their response. We classify these items as CKT-P because they assess a teacher's ability to apply subject-matter knowledge and pedagogical knowledge in support of tasks of teaching. In this context we operationalize a productive response according to the following criteria:

- The teacher must be responding to the disciplinary content of a student statement or idea (Engle and Conant 2002).
- The teacher's response must present an idea or strategy that has a reasonable likelihood of helping the students make progress with their statement or idea.

To be sure, in a real-life interaction, a potentially promising first response to *any* student utterance might be "Tell me more" or "What do you mean by that?" Given that the same response could be given in *any* context and in *any* subject-matter domain, we chose not to count this as a complete answer for the purposes of an energy-specific assessment of CKT in physics.

Atwood's, CKT-P, CR, shown previously in Fig. 2.2, is a constructed-response item that involves responding to a student's question about the energetics of an Atwood's machine. Atwood's, CKT-P, CR is preceded by Atwood's, CKT-D, SR, which prompts teachers to decide whether the total energy of the large block-earth system is increasing, decreasing, or remaining roughly constant. Atwood's, CKT-P, CR is the constructed-response item in which a student raises an insightful question that stems from the need to clarify a system in order to apply work and energy reasoning (see Fig. 2.3). Taylor's statement demonstrates highly metacognitive scientific thinking as she strives to reconcile her understanding of work and gravitational energy. Specifically, Taylor is primed to recognize that earth must be included in the system for the system to have significant gravitational energy. Earth can only do work on the system if it is not included in the system. This is an instantiation of a subtle yet fundamental idea in system-based work and energy analysis. Most high school and college physics teachers introduce the concept of work, so it is very likely that thoughtful students will need to work out this subtle distinction at some point.

We have chosen to present *Atwood's, CKT-P, CR* in this chapter because it represents a difficulty that thoughtful students will often raise when trying to reconcile the concepts of gravitational energy and work by a gravitational force. *Atwood's, CKT-P, CR* is also an item that was relatively difficult for the majority of teachers who participated in our study. We evaluated constructed responses to *Atwood's, CKT-P, CR* on a 3-point rubric in which teachers were awarded one point for productively responding to each component of the item. Productively responding to Taylor's question about work and gravitational energy teaching was relatively difficult for the teachers in our study (see Fig. 2.4). Only 29% of all teachers completing the field test were able to respond productively to any component of the item. A select group of teachers provided answers that addressed all three questions as illustrated by the following examples:

> Taylor needs to understand that work is only done by external forces. I would ask Taylor to reiterate what makes up the system. If she answers that it is the larger block and the earth, I would ask her to remember what kinds of forces are necessary to do work on the system. If she answers that the system is only the large block, then I'd ask her how any gravitational energy could be stored in a system not including the earth.

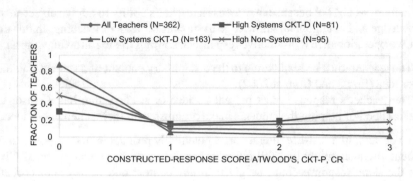

Fig. 2.4 The fraction of teachers who responded productively to student reasoning in *Atwood's, CKT-P, CR*

> a.) A system cannot do work on itself. We already established that the system includes the block and the earth. Therefore, we can't count the work done by gravity because the earth is in the system. b.) What objects did you include in your system? c.) Hopefully she will answer the earth and the block. Then I would ask "Is it possible for an object in the system to do work on the system?" I always relate it back to an aquarium. The fish can do work on each other and the objects inside the fish tank. But the fish can't move the actual tank because they are inside.

We might suggest that the fish tank analogy is problematic because it suggests that the physical configuration, rather than the physicist, determines the system. Nevertheless, both of these teachers correctly interpret the likely inconsistency with which Taylor is struggling. The second teacher also provides evidence that she recognizes this as a prevalent inconsistency that learners encounter.

The majority of teachers were unable to respond productively to any component of the item as illustrated by this example:

> a. Taylor's inconsistency is that she is not thinking of work as a force multiplied by a distance. The work done by gravity comes from the force of gravity (the weight) acting on the box, not on the "gravitational energy" of the block and the Earth. b. I would ask Taylor to define work and describe how it can be calculated. c. If she is able to see that work is the product of a force and a displacement, she would be able to understand that both gravity and the string are doing work, but acting in opposite directions.

The physics content of this response is correct, but it does not address the question Taylor has raised. There is no evidence in Taylor's question to suggest that she does not recognize how work is calculated. In addition, Taylor has explicitly stated that, "both gravity and the string are doing work." Because the suggested instructional response does not address Taylor's question, it is unlikely that it would play a significant role in helping her make progress. Some of the teachers were transparent about their own difficulty interpreting the scientific content at the root of Taylor's question:

> I don't really understand what she is saying when she says "the gravitational energy of the block and the earth." Does she mean the gravitational force between them, like from $F=GmM/r^2$? Or is she referring to the smaller block? I can't really answer because I'm not clear on what she is saying.

2.4.3   Contingency of Productive Attentiveness on Systems Knowledge

In order to respond to the disciplinary content of Taylor's statement, we might expect that a teacher would need a deep understanding of the way in which the choice of system affects work and various potential energies. This expectation was confirmed by comparing responses on *Atwood's, CKT-P, CR* for teachers with low and high systems reasoning composite CKT-D scores. Teachers with low systems CKT-D demonstrated very little success navigating this teaching situation. Only 11% of these teachers were able to respond productively to any portion of the constructed-response items as shown in Fig. 2.4. In contrast, teachers with high systems CKT-D had significantly more success; 69% of them were able to respond productively to some portion of the constructed-response items.[2]

Is it possible that the teachers who responded correctly to the *Atwood's, CKT-P, CR* item just know physics better in general and thus do better on system-based items? Or maybe systems subject-matter knowledge is a separate aspect of energy knowledge, and one can know lots about energy but without understanding systems cannot respond productively to student difficulties of this sort. To test these two explanations, we selected a subgroup of teachers ($N = 95$) based on high scores on nonsystem-related items. The selection criteria for these teachers were entirely independent of their performance on the items assessing systems content knowledge. We call them nonsystems items high-performing teachers. If the first explanation above were correct, then the nonsystems items high-performing teachers would do as well on *Atwood's, CKT-P, CR* as the teachers with a high systems score. If the second one were correct, then the performance of nonsystems items high-performing teachers should be significantly lower than the performance of the teachers with high systems knowledge. We found that out of the nonsystems items high-performing teachers, 49% were able to respond productively to some portion of the constructed-response items compared to 69% of the teachers with high systems knowledge. The difference in performance between each of these two groups was significant at the 0.01 level.

2.5   Patterns in Teacher Quantitative Energy Reasoning

In the preceding section, we presented an example in which attending productively to student reasoning was highly contingent on supporting disciplinary knowledge. This result is not surprising. One might assume that teachers will always need robust

---

[2] *Atwood's, CKT-D, SR* immediately follows *Atwood's CKT-P, CR* and asks subjects to determine if the energy of the large block-earth system is increasing, decreasing, or remaining approximately constant. There are multiple ways to correctly answer this item, some of which do not require a deep understanding of systems reasoning. In contrast, a subject is unlikely to answer all parts of *Cyclist, CKT-D, SR* correctly without a deep understanding of systems reasoning. This was our motivation for using a composite systems CKT-D score that provides a clear gauge of a teacher's disciplinary knowledge for systems reasoning.

Two students in Ms. Engel's physics class are discussing the energetics of dribbling a basketball on a wooden floor. They agree that all of the kinetic energy gets converted into elastic energy for an instant when the basketball is compressed the most. They also agree that many objects, even basketballs and wooden floors, can be modeled as springs. They are uncertain about whether there would be equal amounts of elastic energy in the ball and the floor. They call Ms. Engel over to share their ideas with her and get some help.

Marcos says, "We were thinking that when the ball compresses against the floor, the forces that the ball and the floor exert on each other would be equal and opposite, so maybe the amount of elastic energy in the floor is the same as the elastic energy in the ball."

Louisa responds, "I get that the forces are the same, but I am thinking that the ball compresses more than the floor, so shouldn't there be more energy stored in the ball?"

Marcos replies, "But the floor is more rigid and would have a higher spring constant. I think the larger k of the floor compensates for the smaller Δx in the $\frac{1}{2}k(\Delta x)^2$ equation, and the elastic energies are the same."

1. Is Marcos correct that the elastic energy of the ball and the floor would be the same?
   A. Yes. (31%)
   B. No. The elastic energy of the ball would be greater. (51%)
   C. No. The elastic energy of the floor would be greater. (1%)
   D. There is not enough information to compare these energies. (17%)

2. Which of the following activities would be most likely to provide Ms. Engel's students with additional insights about the relative amounts of elastic energy during the bounce of the basketball?
   A. They could measure the spring constants and the displacements of both the floor and the ball and use those to compare the elastic energies. (10%)
   B. They could compare the elastic energies of two non identical springs when they are compressed with the same force. (70%)
   C. They could do an experiment to see if a basketball bounces higher on a soft carpet surface or a hard concrete floor. (14%)
   D. They could do an experiment to show that the same basketball will not bounce as high off the gym floor if it has first been put in a freezer. (6%)

Explain your selection and how the activity you selected might provide the students with additional insights about the relative elastic energy during the bounce of a basketball

Fig. 2.5 *Basketball, CKT-D, SR and Basketball, CKT-P, CR questions*

disciplinary knowledge in order to respond productively to the scientific thinking of their students. One might even assume that teachers themselves must know the correct answer in order to help their students make progress toward that answer. In this section, we present an example from quantitative energy reasoning that complicates the first assumption and challenges the second.

2.5.1 Attending Productively to Quantitative Energy Reasoning

Basketball, CKT-P, SR and *CR* in Fig. 2.5 are selected-response and constructed-response questions that involve responding to a student dialogue about the energetics of a bouncing basketball. The classroom scenario described in this item was inspired by real classroom experiences. Basketball, *CKT-P, SR* and *Basketball, CKT-P, CR* both assess a teacher's ability to identify an instructional activity that would allow the students to build on and refine their quantitative energy reasoning. The student statements provide some of the essential components of a mathematical approach to this item. For the same exerted force, the compression of the object is inversely proportional to the effective spring constant. The resulting elastic energy will then be greater for the object with the lower spring constant and a corresponding greater amount of compression.

This item requires knowledge that is beyond the domain of pure subject-matter knowledge. The teacher should recognize additional ideas that could help the students resolve their debate and identify a specific activity, in this case a specific experiment that could be feasibly carried out in the classroom and would allow the students to develop these ideas further. Specifically, the students could use two springs with different spring constants, compress them exerting the same force, measure the compressions, and compare the energy of two nonidentical springs when they are compressed with the same force. This approach would allow the students to collect data to mathematically compare the elastic energy in the simpler, and more familiar, example of two springs. Once they have worked out the energy comparison for two simple springs, they should be able to apply this comparative model to the basketball bouncing on the floor. This item includes both a selected-response portion and a constructed-response portion, the latter of which invites the teacher to explain her choice. Scoring rubrics for the constructed-response answers were developed and refined to achieve an inter-rater reliability at least as large as 90%.

Seventy percent of teachers in our sample selected the most productive instructional response from the choices provided on *Basketball, CKT-P, SR*. Seventy-three percent of teachers provided at least a partially correct explanation of how their chosen instructional response could provide students with additional insights about the energetics of a bouncing basketball as shown in Fig. 2.6. The following examples both illustrate full explanations of the most productive instructional response:

Their argument seems to hinge on the difference between the linear relationship in $F=-kx$ and the squared relationship in energies. By looking at two springs with measurable and quantifiable spring constants and compressions, they can see that while forces may be

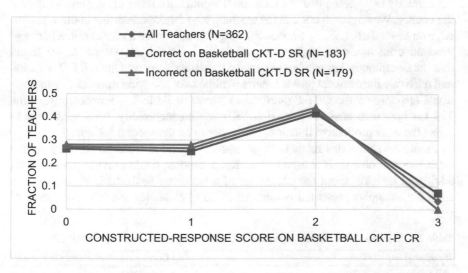

Fig. 2.6 The fraction of teachers who correctly explained in the *CR* question their responses to the *Basketball, CKT-P, SR* question

equal, relative energies are dominated by the squared term. The other experiments are useful to show how modification to one or the other changes things, but their argument seems to hinge on this mathematical confusion which should first be analyzed with more simplistic and easily quantifiable things. [...]

Since the floor and the ball aren't identical, but do experience the same force, the chosen scenario is the easiest to actually do an experiment with. If the same force yields the same stored energy for each spring, the ball and the floor would store the same amount of energy. If the experiment shows that the two springs store different amounts of energy, then the ball and the floor store different amounts. I haven't done the experiment, but I would expect the ball to be the weaker spring and therefore store more energy because of the x^2 in the PE function.

Why were teachers so much more successful in responding productively to this teaching situation than in the case of the Atwood's example? Perhaps the relevant CKT-D of quantitative energy reason is simply much more widely held than the CKT-D, which supports systems reasoning.

2.5.2 Contingency of Productive Attentiveness on Quantitative Energy Knowledge

Based on our findings for the *Atwood's* item, we might expect that a teacher would be better prepared to suggest a productive strategy for comparing the elastic energies of the basketball and the floor if they had a correct understanding of how these elastic energies actually do compare. *Basketball, CKT-D, SR* provides teachers with the opportunity to demonstrate this understanding. Slightly more than half of teachers correctly recognized that the basketball would have more elastic energy during the bounce. We might expect that the teachers who displayed correct content knowledge on *Basketball, CKT-D, SR* would be significantly more likely to select the most productive instructional response on *Basketball, CKT-P, SR*. However, we found that the disciplinary knowledge necessary to correctly answer the *CKT-D* question had relatively little impact on a teacher's likelihood of selecting a productive instructional response on the *CKT-P* question as shown in Table 2.4. Surprisingly, of the 179 teachers who answered the first SR question incorrectly, 66% were able to select the most productive instructional response on the second SR question.

Figure 2.6 shows that many of these teachers provided at least a partially correct explanation of how their chosen instructional response could provide students with additional insights about the energetics of a bouncing basketball, as one can see from the examples presented below. All of these examples are from the group of

Table 2.4 Teachers' performance on *Basketball, CKT-D, SR* and *Basketball, CKT-P, SR*

Teacher group	Correct on *Basketball, CKT-P, SR*
All teachers ($N = 362$)	255 (70%)
Correct on *Basketball, CKT-D, SR* ($N = 183$)	136 (74%)
Incorrect on Basketball, CKT-D, SR ($N = 179$)	119 (66%)

teachers who incorrectly agreed with Carlos that the ball and the floor would have the same elastic potential energies.

> The idea that the gym floor has a higher k value because it is more stiff than the ball should be obvious to most. Measuring the spring constants of the floor and the ball are probably not feasible in a high school science class, otherwise the first choice would have been the best. The forces are equal based on Newton's 3rd Law. I thought number 2 was the best choice because you could see how different k values, but identical forces affect the elastic energies of the springs, but it would also be practical to do in a school.
>
> Although I like the idea of testing Marcos's hypothesis in option one, the difficulty of finding the displacement of the floor and the very real possibility of having a particular trial happen to have the same elastic energy is a problem. By testing two non-identical springs and the resulting energy from not only the same force but different identical forces, will help eliminate variables out of your control while still addressing the issue of two different spring constants interacting with each other like the ball and the floor.

Surprisingly, the disciplinary knowledge necessary to correctly answer *Basketball, CKT-D, SR* had relatively little impact on a teacher's propensity to select or explain a productive instructional response. The small difference in performance on *Basketball, CKT-P, SR* among the teachers who selected the correct CKT-D response (74%) versus the teachers who did not (66%) was not significant at the 0.05 level. The performance on *Basketball, CKT-P, CR* was also similar.

2.6   Summary and Implications for Instruction

In this paper we set out to answer the following research questions in the context of a paper-and-pencil assessment:

1. To what degree is the productivity of teacher responses to student thinking contingent on the teacher's content knowledge of the relevant physics topic?
2. Are there specific aspects of content knowledge of the relevant topic that are critical for supporting student thinking?

2.6.1   Identifying Foundational and Elaborative CKT-D

Our findings suggest that in some cases, responding productively to students is contingent on strong supporting disciplinary knowledge. This may be especially true when disciplinary knowledge relates to a fundamental or axiomatic rather than experimentally testable aspect of the energy model.[3] In our study such a fundamental aspect of the energy model is the concept of a system. In physics, the underlying idea of a system is that there is a choice of objects to be included in the system, but once the choice is made, these objects cannot exchange energy with the system. Their interactions and motions contribute to the system's total energy. Only objects that are external to the system can exchange energy with the system. One can write equations

---

[3] There is no experiment that one could conduct to decide how one should select a system.

for energy types and solve simple problems involving such calculations without a deep understanding of this idea; however, it is often not possible to track energy flow and make real connections between energy and mechanisms for energy transfer without an understanding of this foundational piece of the energy framework. A good example here is knowing when to apply the work-kinetic energy theorem (the work done by the net force exerted on an object is equal to the change of the object's kinetic energy) or a more general energy conservation statement (the change of the energy of the system is equal to the work done by external forces on the system assuming that thermal energy transfer is zero, as it is in our case). Here the subtle difference between a point particle model in the first case and an arbitrarily chosen system of interacting objects in the second case is crucial. Experts understand this difference and do not confuse the two, while novices often do (Arons 1999; Sherwood 1983). Can a teacher help a student recognize the difference if she does not have a solid understanding of the concept of a system? Based on our study, we answer this question in the negative. The reason here lies in the fundamental, model-based nature of system reasoning in physics, which does not lend itself to experimental tests and yet is the foundation of all further investigations.

However, we also found that in some cases, responding productively to student thinking is not contingent on strong subject-matter knowledge. This may be especially true when this kind of knowledge supports a specific elaboration (application) of the energy model. In our study we found that even those teachers who were not successful in applying the mathematical expression for elastic potential energy could still choose a productive teaching approach to help their students understand the difference between force and energy. This approach involved deciding what experiment students could conduct to test their ideas. We found that when the teachers had an option to choose a strategy to help a student understand the nature of the mathematical expression for elastic potential energy, more teachers than in the case of systems were successful in choosing a productive experiment.

2.6.2   Implications for Teacher Professional Preparation

In a traditional approach to teacher preparation, it is assumed that prospective teachers develop content knowledge (everything they need to know about the discipline), while they are taking courses in the science departments and later learn teaching-related knowledge in the schools of education. Research into PCK, and more recently CKT, provides frameworks for understanding the limitations of this traditional approach. The disciplinary knowledge that physics teachers draw on in their classrooms extends beyond the content knowledge of physics majors. Thus, if we wish to prepare and graduate qualified teachers, they need to develop content knowledge for teaching their respective disciplines (Ball et al. 2008; Ball 2000). This means that we need to develop special methods courses where future teachers learn how to teach their subject matter (e.g., physics, chemistry, biology, mathematics) instead of putting them all together in one generic science methods course

(Etkina 2010). Development of such courses is challenging and should be informed by formative assessment of discipline-specific content knowledge for teaching (Bailie 2017). It is critical to prioritize knowledge and strategies that will best support teachers in responding productively to student thinking. We believe that this study presents a model for identifying specific areas of disciplinary knowledge that are both essential and often deficient among practicing teachers.

The disciplinary knowledge base that teachers deploy to support and respond productively to student thinking is subtle, complex, and extensive. Therefore, we should hope that the requisite knowledge base will be constructed and refined through reflective experience in the classroom. Our results provide both encouraging and cautionary insights into the in-service development of CKT. We are encouraged by cases where teachers are able to utilize their knowledge of pedagogy and scientific practices to compensate for a lack of CKT-D. This provides a promising avenue for continued growth of disciplinary knowledge throughout a teacher's career. In order to prime teachers for this type of ongoing professional growth, teacher preparation programs should emphasize domain-specific pedagogical and scientific practices. Our results also suggest that a greater emphasis on pedagogy and practices is necessary but not sufficient. Even a robust understanding of these practices will not compensate for foundational or axiomatic disciplinary knowledge that cannot be tested empirically, such as the physics-specific relationship of work and systems.

It is very important that those who prepare teachers in a specific content area are in agreement regarding what constitutes foundational content knowledge in their discipline (in the areas relevant to K–12 curriculum) and what science practices are most helpful for responding to student ideas. The Next Generation Science Standards (NGSS) disciplinary core ideas, crosscutting concepts, and science practices provide a framework for enumerating these foundational ideas and practices, but we feel that additional evidence-based refinement is necessary. For example, in physics one of the practices that helps students test their ideas or that helps teachers respond productively to student ideas is follow-up of data analysis and interpretation of experimentation (whether real or imaginary). Additional attention should be given to this aspect of physics teacher preparation, but an increased emphasis on traditional laboratories will not empower teachers to respond productively to student thinking. We envision modeling the design of experiments to responsively test learner ideas as an integral and extensive component of teacher professional preparation programs. It is unrealistic to expect teachers to establish empirically responsive classrooms if they have not previously participated in such a learning community.

Acknowledgments We are grateful to the present and former project members at Rutgers University (Candice Dias Drew Gitomer, Charles Iaconangelo and Robert Zisk); Seattle Pacific University (Abigail Daane, Lezlie Salvatore DeWater, Lisa Goodhew, Kara Gray, Amy Robertson, Hannah Sabo, Rachel Scherr, and Orlala Wentink); Educational Testing Service (Courtney Bell, Geoffrey Phelps, and Barbara Weren); Facet Innovations, LLC (Ruth Anderson and Jim Minstrell); Horizon Research, Inc. (Sean Smith); and the University of Maine (Michael Wittmann). We are particularly grateful for the Leanna Akers who assisted with data analysis along with Orlala Wentink, Colleen McDermott, and Courtney Bell who provided editorial assistance on this manuscript. Finally, we thank the National Science Foundation for its support of this work (DRL Award number 1222777).

Appendices

Appendix A: Tasks of Teaching

Task of teaching	Description	Specific tasks
I. Anticipating student thinking around science ideas	*While planning and implementing instruction, teachers are able to anticipate particular patterns in student thinking. They understand and recognize challenges students are likely to confront in developing an understanding of key science concepts and mathematical models. Teachers are also familiar with student interests and background knowledge and enact instruction accordingly*	*Teachers* Ia. Anticipate specific student challenges related to constructing scientific concepts, conceptual and quantitative reasoning, experimentation, and the application of science processes Ib. Anticipate likely partial conceptions and alternate conceptions, including partial quantitative understanding about particular science content and processes Ic. Recognize student interest and motivation around particular science content and practices Id. Understand how students' background knowledge both in physics and mathematics can interact with new science content

(continued)

Appendix A   (continued)

Task of teaching	Description	Specific tasks
II. Designing, selecting, and sequencing learning experiences and activities	*Classroom learning experiences and activities are designed around learning goals and involve key science ideas, key experiments, and mathematical models relevant to the development of ideas and practices. Learning experiences reflect an awareness of student learning trajectories and support both individual and collective knowledge generation on the part of students*	*Teachers* IIa. Design or select sequence learning experiences that focus on sense-making around important science concepts and practices, including productive representations, mathematical models, and experiments in science that are connected to students' initial and developing ideas IIb. Include key practices of science including experimentation, reasoning based on collected evidence, experimental testing of hypotheses, mathematical modeling, representational consistency, and argumentation IIc. Address projected learning trajectories that include both long-term and short-term goals and are based on evidence of actual student learning trajectories IId. Address learners' actual learning trajectories by building on productive elements and addressing problematic ones IIe. Provide students with evidence to support their understanding of short- and long-term learning goals IIf. Integrate, synthesize, and use multiple strategies and involve students in making decisions IIg. Prompt students to collectively generate and validate knowledge with others IIh. Help students draw on multiple types of knowledge, including declarative, procedural, schematic, and strategic IIi. Elicit student understanding and help them express their thinking via multiple modes of representation IIj. Help students consider multiple alternative approaches or solutions, including those that could be considered to be incorrect

(continued)

Appendix A (continued)

Task of teaching	Description	Specific tasks
III. Monitoring, interpreting, and acting on student thinking	*Teachers understand and recognize challenges and difficulties students experience in developing an understanding of key science concepts; understanding and applying mathematical models and manipulating equations; designing and conducting experiments,* etc. *This is evident in classroom work, talk, actions, and interactions throughout the course of instruction so that specific learning needs or patterns are revealed Teachers also recognize productive developing ideas and problem solutions and know how to leverage these to advance learning* *Teachers engage in an ongoing and multifaceted process of assessment, using a variety of tools and methods. Teachers draw on their understanding of learners and learning trajectories to accurately interpret and productively respond to their students' developing understanding*	*Teachers* IIIa. Employ multiple strategies and tools to make student thinking visible IIIb. Interpret productive and problematic aspects of student thinking and mathematical reasoning IIIc. Identify specific cognitive and experiential needs or patterns of needs and build upon them through instruction IIId. Use interpretations of student thinking to support instructional choices both in lesson design and during the course of classroom instruction IIIe. Provide students with descriptive feedback IIIf. Engage students in metacognition and epistemic cognition IIIg. Devise assessment activities that match their goals of instruction

(continued)

Appendix A (continued)

Task of teaching	Description	Specific tasks
IV. Scaffolding meaningful engagement in a science learning community	*Productive classroom learning environments are community-centered. Teachers engage all students as full and active classroom participants. Knowledge is constructed both individually and collectively, with an emphasis on coming to know through the practices of science. The values of the classroom community include evidence-based reasoning, the pursuit of multiple or alternative approaches or solutions, and the respectful challenging of ideas*	*Teachers* IVa. Engage all students to express their thinking about key science ideas and encourage students to take responsibility for building their understanding, including knowing how they know IVb. Develop a climate of respect for scientific inquiry and encourage students' productive deep questions and rich student discourse IVc. Establish and maintain a "culture of physics learning" that scaffolds productive and supportive interactions between and among learners IVd. Encourage broad participation to ensure that no individual students or groups are marginalized in the classroom IVe. Promote negotiation of shared understanding of forms, concepts, mathematical models, experiments, etc., within the class IVf. Model and scaffold goal behaviors, values, and practices aligned with those of scientific communities IVg. Make explicit distinctions between science practices and those of everyday informal reasoning as well as between scientific expression and everyday language and terms IVh. Help students make connections between their collective thinking and that of scientists and science communities IVi. Scaffold learner flexibility and the development of independence IVj. Create opportunities for students to use science ideas and practices to engage real-world problems in their own contexts

(continued)

Appendix A (continued)

Task of teaching	Description	Specific tasks
V. Explaining and using examples, models, representations, and arguments to support students' scientific understanding	*Teachers explain and use representations, examples, and models to help students develop their own scientific understanding. Teachers also support and scaffold students' ability to use models, examples, and representations to develop explanations and arguments. Mathematical models are included as a key aspect of physics understanding and are assumed whenever the term* model *is used*	*Teachers* Va. Explain concepts clearly, using accurate and appropriate technical language, consistent multiple representations, and mathematical representations when necessary Vb. Use representations, examples, and models that are consistent with each other and with the theoretical approach to the concept that they want students to learn Vc. Help students understand the purpose of a particular representation, example, or model and how to integrate new representations, examples, or models with those they already know Vd. Encourage students to invent and develop examples, models, and representations that support relevant learning goals Ve. Encourage students to explain features of representations and models (their own and others') and to identify/evaluate both strengths and limitations Vf. Encourage students to create, critique, and shift between representations and models with the goal of seeking consistency between and among different representations and models Vg. Model scientific approaches to explanation, argument, and mathematical derivation and explain how they know what they know. They choose models and analogs that accurately depict and do not distort the true meaning of the physical law and use language that does not confound technical and everyday terms (e.g., heat and energy). Vh. Provide examples that allow students to analyze situations from different frameworks such as energy, forces, momentum, and fields

(continued)

Appendix A   (continued)

Task of teaching	Description	Specific tasks
VI. Using experiments to construct, test, and apply concepts	*Teachers provide timely and meaningful opportunities throughout instruction for students to design and analyze experiments to help students develop, test, and apply particular concepts. Experiments are an integral part of student construction of physics concepts and are used as part of scientific inquiry in contrast with simple verification*	*Teachers* VIa. Provide opportunities for students to analyze quantitative and qualitative experimental data to identify patterns and construct concepts VIb. Provide opportunities for students to design and analyze experiments using particular frameworks such as energy, forces, momentum, field, etc. VIc. Provide opportunities for students to test experimentally or apply particular ideas in multiple contexts IVd. Provide opportunities for students to pose their own questions and investigate them experimentally VIe. Use questioning, discussion, and other methods to draw student attention during experiments to key aspects needed for subsequent learning, including the limitations of the models used to explain a particular experiment VIf. Help students draw connections between classroom experiments, their own ideas, and key science ideas VIg. Encourage students to draw on experiments as evidence to support explanations and claims and to test explanations and claims by designing experiments to rule them out

Appendix B: Energy-Specific Student Targets

(Energy-Related Content and Practice Ideas)

A. Connections of energy and everyday experiences
    The student:

1. Uses energy ideas to interpret or explain everyday phenomena
2. Recognizes the important role of internal energy in interpreting or explaining everyday phenomena

B. Choice of system
    The student:

1. Recognizes that the energy accounting in a phenomenon depends on the choice of system
2. Explains the relative advantage of a given system choice (i.e., relative ease of analysis)
3. Recognizes that the choice of system determines whether springs or earth do work (i.e., if the spring or earth are in the system, they do not do any work on the system, but the system can possess elastic or gravitational potential energy)
4. Identifies and differentiates between forms of energy and other physics concepts

C. Identification of and differentiation between forms of energy and other physics concepts
    The student:

1. Recognizes that energy cannot be observed directly and knows how different forms of energy correspond to different measurable physical quantities
2. Recognizes and maintains a consistency of scale (microscopic or macroscopic) during energy analysis
3. Differentiates between energy and related ideas (e.g., force, power, stimulus, trigger, activation, speed, distance, temperature)
4. Distinguishes between forms of energy and energy transfers

D. Transfer of energy (environment → system; system → environment)
    The student:

1. Recognizes that the energy of a system is always conserved but might not be constant
2. Recognizes that work is the way in which energy is transferred mechanically and may result in a change in temperature in some cases
3. Avoids double counting when analyzing processes involving work and energy
4. Recognizes when to use compensatory models for tracking energy into and out of a system and when quantitative models are of limited use

E. Use of mathematics
   The student:

1. Understands that when considering potential energy, it is important to think about the change. The zero level of potential energy is arbitrary, but the change is not. The energy of attraction is negative if the zero level is set at infinity.
2. Can account for vector and scalar quantities in energy analysis.
3. Understands that work is a scalar quantity, and the positive or negative sign of work does not indicate direction but addition or subtraction.
4. Connects forms of energy and the factors on which they depend through appropriate linear and nonlinear mathematical relationships.
5. Applies conservation as a mathematical constraint on the outcomes of possible processes.
6. Recognizes that the mathematical analysis of energy-related processes depends on the choice of initial and final state and the choice of system.

F. Use of representations
   The student:

1. Selects/creates and uses appropriate verbal, mathematical, and graphical/pictorial representations (specific for energy, such as bar charts, energy diagrams, etc.) to describe, analyze, and/or communicate a physical situation or process
2. Interprets different representations used to describe, analyze, and/or communicate a physical situation or process
3. Understands the relationships between different representations of the same phenomenon and seeks consistency among different representations
4. Understands standard technical representations and language used to communicate energy-related ideas

G. Use of science practices
The student:

1. Uses a range of representations to communicate ideas and illustrate or defend explanations

2. Connects energy ideas to other learning and real-life processes and projects through experimental investigations, energy problem solutions, and engineering designs

3. Designs experiments to test competing hypotheses

4. Makes choices in data collection and analysis that allow for inferring the amounts and transfers of energy even when they cannot be measured directly

5. Connects experiments and data to the mathematical representations of energy

6. Evaluates and negotiates choices/options by considering the merits, limitations, and relative advantages of different engineering designs in terms of, for example, different choices of energy models for the same physical process

7. Provides evidence-based arguments concerning energy processes and engineering designs

8. Demonstrates consistency and coherence in model-based and evidence-based reasoning in making predictions and interpreting results

References

Arons, A. B. (1999). Development of energy concepts in introductory physics courses. *American Journal of Physics, 67*(12), 1063.

Bailie, A. L. (2017). Developing preservice secondary science teachers' pedagogical content knowledge through subject area methods courses: A content analysis. *Journal of Science Teacher Education, 28*(7), 631–649 (2017).

Ball, D. L. (2000). Bridging practices: Intertwining content and pedagogy in teaching and learning to teach. *Journal of Teacher Education, 51*(3), 241–247.

Ball, D. L., Thames, M. H., & Phelps, G. (2008). Content knowledge for teaching: What makes it special? *Journal of Teacher Education, 59*(5), 389–407.

Berry, A., Friedrichsen, P., & Loughran, J. (Eds.). (2015). *Re-examining pedagogical content knowledge in science education*. New York, NY: Routledge.

Besson, U., & Viennot, L. (2004). Using models at the mesoscopic scale in teaching physics: Two experimental interventions in solid friction and fluid statics. *International Journal of Science Education, 26*, 1083–1110.

Brewe, E. (2011). Energy as a substancelike quantity that flows: Theoretical considerations and pedagogical consequences. *Physical Review Special Topics – Physics Education Research, 7*(2), 1–14.

Daane, A. R., Vokos, S., & Scherr, R. E. (2013). Learner understanding of energy degradation. In *2013 physics education research conference*.

Daane, A. R., Vokos, S., & Scherr, R. E. (2014). Goals for teacher learning about energy degradation and usefulness. *Physical Review Special Topics – Physics Education Research, 10*(2), 1–16.

Daane, A. R., McKagan, S. B., Vokos, S., & Scherr, R. E. (2015). Energy conservation in dissipative processes: Teacher expectations and strategies associated with imperceptible thermal energy. *Physical Review Special Topics – Physics Education Research, 11*(1), 1–15.

Engle, R. A., & Conant, F. R. (2002). Guiding principles for fostering productive disciplinary engagement: Explaining an emergent argument in a community of learners classroom. *Cognition and Instruction, 20*(4), 399–483.

Etkina, E. (2010). Pedagogical content knowledge and preparation of high school physics teachers. *Physical Review Special Topics – Physics Education Research, 6*, 1–26.

Etkina, E. (2015). Millikan award lecture: Students of physics—listeners, observers, or collaborative participants in physics scientific practices? *American Journal of Physics, 83*, 669.

Etkina, E., Gitomer, D., Iaconangelo, C., Phelps, G., Seeley, L., & Vokos, S. (2018). Design of an assessment to probe teachers' content knowledge for teaching: An example from energy in HS physics. *Physical Review Physics Education Research, 14*, 1–20.

Gess-Newsome, J. (2011). Pedagogical content knowledge. In J. Hattie & E. Anderman (Eds.), *International handbook of student achievement*. New York, NY: Routledge.

Gess-Newsome, J. (2015). A model of teacher professional knowledge and skill including PCK: Results of the thinking from PCK summit. In A. Berry, P. Friedrichsen, & J. Loughran (Eds.), *Re-examining pedagogical content knowledge in science education*. New York, NY: Routledge.

Gitomer, D., Phelps, G., Weren, B., Howell, H., & Croft, A. (2014). Evidence on the validity of content knowledge for teaching assessments. In T. J. Kane, K. A. Kerr, & R. C. Pianta (Eds.), *Designing teacher evaluation systems: New guidance from the measures of effective teaching project*. San Francisco, CA: Jossey-Bass.

Keller, M. M., Neumann, K., & Fischer, H. E. (2017). The impact of physics teachers' pedagogical content knowledge and motivation on students' achievement and interest. *Journal of Research in Science Teaching, 54*(5), 586–614.

Lindsey, B. A. (2014). Student reasoning about electrostatic and gravitational potential energy: An exploratory study with interdisciplinary consequences. *Physical Review Special Topics – Physics Education Research, 10*(1), 1–6.

Lindsey, B. A., Heron, P. R. L., & Shaffer, P. S. (2009). Student ability to apply the concepts of work and energy to extended systems. *American Journal of Physics, 77*(11), 999.

Lindsey, B. A., Heron, P. R. L., & Shaffer, P. S. (2012). Student understanding of energy: Difficulties related to systems. *American Journal of Physics, 80*(2), 154.

Magnusson, S., Krajcik, J., & Borko, H. (1999). Nature, sources, and development of pedagogical content knowledge for science teaching. In J. Gess-Newsome & N. G. Lederman (Eds.), *Examining pedagogical content knowledge: The construct and its implications for science education*. Dordrecht, The Netherlands: Kluwer Academic Publishers.

National Research Council. (1999). *How people learn: Bridging research and practice*. Washington, DC: The National Academies Press.

National Research Council. (2000). *How people learn: Brain, mind, experiences and school*. Washington, DC: The National Academies Press. Expanded edition.

National Research Council. (2012). *A framework for K–12 science education: Practices, crosscutting concepts, and core ideas*. Washington, DC: The National Academies Press.

Phelps, G., Gitomer, D. H., Iaconangelo, C. J., Etkina, E., Seeley, L., & Vokos, S. (in preparation). *Assessing the content knowledge used in teaching energy*.

Robertson, A. D., Scherr, R. E., & Hammer, D. (Eds.). (2016). *Responsive teaching in science and mathematics*. New York, NY: Routledge.

Scherr, R. E., Close, H. G., McKagan, S. B., & Vokos, S. (2012a). Representing energy. I. Representing a substance ontology for energy. *Physical Review Special Topics – Physics Education Research, 8*(2), 1–11.

Scherr, R. E., Close, H. G., McKagan, S. B., & Vokos, S. (2012b). Representing energy. II. Energy tracking representations. *Physical Review Special Topics – Physics Education Research, 8*(2), 1–11.

Scherr, R. E., Robertson, A. D., Seeley, L., & Vokos, S. (2013a). Content knowledge for teaching energy: An example from middle-school physical science. In *2012 physics education research conference*.

Scherr, R. E., Close, H. G., Close, E. W., Flood, V. J., McKagan, S. B., Robertson, A. D., et al. (2013b). Negotiating energy dynamics through embodied action in a materially structured environment. *Physical Review Special Topics – Physics Education Research, 9*(2), 1–18.

Seeley, L., Vokos, S., & Minstrell, J. (2014). Constructing a sustainable foundation for thinking and learning about energy in the twenty-first century. In R. F. Chen, A. Eisenkraft, D. Fortus, J. Krajcik, K. Neumann, J. Nordine, & A. Schef (Eds.), *Teaching and learning of energy in K–12 education* (pp. 337–356). New York, NY: Springer International Publishing.

Sherwood, B. A. (1983). Pseudowork and real work. *American Journal of Physics, 51*(7), 597.

Shulman, L. S. (1986). Those who understand: Knowledge growth in teaching. *Educational Research, 15*(2), 4–14.

Shulman, L. S. (1987). Knowledge and teaching: Foundations of the new reform. *Harvard Educational Review, 57*, 1–22.

Swackhamer, G. (2005). *Cognitive resources for understanding energy* [Online]. Available: http://
   modeling.asu.edu/modeling/CognitiveResources-Energy.pdf. Accessed 24 May 2017.
Van Heuvelen, A., & Zou, X. (2001). Multiple representations of work–energy processes. *American
   Journal of Physics, 69*(2), 184.
Vollmer, M., & Mollmann, K.-P. (2010). *Infrared thermal imaging: Fundamentals, research and
   applications*. Weinheim, Germany: Wiley-VCH.

Chapter 3
Personal and Canonical PCK: A Synergistic Relationship?

P. Sean Smith, Courtney L. Plumley, Meredith L. Hayes, and R. Keith Esch

Abstract For 30 years, science education researchers and practitioners have waited for the promise of pedagogical content knowledge (PCK) to be fulfilled. PCK has the potential to shape instruction, teacher professional learning, and instructional materials. When the field speaks about PCK in terms of these benefits, a particular kind of PCK is envisioned. In our work, we refer to this kind of PCK as "canonical" to convey that it is widely accepted by the field and transcends context. Despite its promise, examples of canonical PCK are lacking in relation to the number of science topics in standards documents. In this chapter, we explore the possibility that the PCK held by teachers—"personal PCK"—can be compiled to grow the body of canonical PCK. We first describe a model of personal-canonical PCK synergy. We then explain how we have tested this synergy hypothesis, drawing on literature reviews and data collected directly from teachers. We find that, within the narrow range of topics we have focused on, personal PCK does not accumulate to fill gaps in the canon. We illustrate through several examples that instead, personal PCK appears largely as variations on PCK themes already apparent in the literature. We conclude the chapter by discussing implications for the field.

Keywords PCK · Personal PCK · Canonical PCK · Student thinking

3.1 Introduction

The construct of pedagogical content knowledge (PCK) is 30 years old and has an extensive literature on its many facets, including how it is defined, developed, elicited, assessed, and measured. Rather than review this literature comprehensively, we begin by using selected pieces to establish our orientation toward PCK.

P. S. Smith (✉) · C. L. Plumley · M. L. Hayes · R. K. Esch
Horizon Research, Inc., Chapel Hill, NC, USA

© Springer International Publishing AG, part of Springer Nature 2018
S. M. Uzzo et al. (eds.), *Pedagogical Content Knowledge in STEM*,
Advances in STEM Education, https://doi.org/10.1007/978-3-319-97475-0_3

Shulman described PCK as "that special amalgam of content and pedagogy that is uniquely the province of teachers…" (Shulman 1987, p. 8). The word "amalgam" has important implications—among them, both content knowledge and pedagogical knowledge are necessary for PCK; neither is alone sufficient. Our particular orientation toward PCK aligns with that of others who argue that the content dimension is not just domain specific but topic specific as well (e.g., Gess-Newsome 2015; Veal and MaKinster 1999)—e.g., PCK exists for science, for chemistry, and for the topic of equilibrium within chemistry. Our work focuses exclusively on topic-specific PCK. Specifically, in this chapter, we use illustrations from our work on two fifth-grade topics (or disciplinary core ideas) in the Next Generation Science Standards ([NGSS] NGSS Lead States 2013): the Small Particle Model of Matter and Interdependent Relationships in Ecosystems.[1]

In attempts to parse the construct of PCK, several researchers have proposed categories. All researchers seem to agree on two broad categories of topic-specific PCK: knowledge of instructional strategies and knowledge of student thinking. The former may include laboratory activities, simulations, and ways to elicit student thinking, among others. The latter includes prominent misconceptions[2] and learning progressions. We have found these two broad categories particularly helpful in our own work, as have others (e.g., Alonzo and Kim 2016).

Shulman's original conception of PCK clearly describes knowledge that resides in teachers, what some researchers refer to as "personal PCK" (e.g., Gess-Newsome 2015). A critical feature of our orientation is the assertion that PCK can exist external to teachers, available to all in the same way that science knowledge is available to all in books and other forms. We characterize this form of PCK as "canonical" to suggest that it is, like canonical science knowledge, widely accepted by the field. Shulman described a construct similar to canonical PCK when he wrote about the collected and codified "wisdom of practice among both inexperienced and experienced teachers" (Shulman 1987, p. 11). Examples of synthesized canonical PCK are limited, but they do exist and tend to focus on student thinking, in particular the well-known works of Rosalind Driver and colleagues (e.g., Driver 1994; Driver and Easley 1978; Driver et al. 1985). For example, several studies have found that young students often believe liquid substances not only disappear when they evaporate but the matter itself ceases to exist (Lee et al. 1993; Osborne and Cosgrove 1983; Russell et al. 1989; Tytler and Peterson 2000). However, relative to the number of science topics in K–12 standards documents, examples of widely accepted patterns in student thinking are sparse, and examples of widely accepted instructional PCK are even more elusive. We hypothesize that personal and canonical PCK can have a synergistic relationship, which we elaborate on below.

---

[1] Using NGSS notation, these topics correspond to DCIs 5-PS1.A and 5-LS2.A, respectively. We selected these topics because of our focus on upper elementary science instruction and because of the contrast they offer between physical and life science.

[2] We define misconceptions as student ideas that (1) are in conflict with accepted scientific ideas and (2) form through interaction with the natural world. Misconceptions are neither good nor bad, but they do tend to be deeply ingrained in students' thinking. Some are part of a learning progression for a topic, suggesting that many students will have them at some point as they develop full understanding. Examples include (1) air does not have mass, and (2) plants get their food from soil.

3.2   A Model of Personal-Canonical PCK Synergy

A robust PCK canon could benefit the field in important ways. For example, it could form the basis for curriculum materials design, for professional development design, and for PCK assessments. Figure 3.1 shows a model that suggests a synergistic relationship between canonical and personal PCK. The model was first formalized in preparation for an international meeting on science PCK in 2012 (Carlson et al. 2015), in which the lead author of this chapter participated.

The model asserts that canonical PCK exists and that it traditionally emerges from research on student thinking and instructional strategies. However, with the exception of Driver's work on student thinking referenced above, efforts to synthesize such research have been infrequent. Personal PCK, as described in the model, forms through teaching (or teaching-related) experience. For example, PCK may develop as a teacher plans for instruction or enacts a lesson and monitors its effect on students. We have found personal PCK particularly difficult to elicit from teachers, as we describe later in this chapter.

The model suggests that it is possible, but not inevitable (as represented by the dashed lines), for a synergistic relationship to exist between canonical and personal PCK; however, neither is dependent on the other. The first aspect of synergy is evident in the assertion that personal PCK, through consensus among many teachers, may ultimately become canonical PCK. We have been testing this hypothesis in our work and summarize our findings later in this chapter. Efforts to collect and make public the consensus of many teachers' personal PCK are not widespread, but there are notable precedents. Researchers at Monash University in Australia pioneered such work on a local scale through their CoRe instrument (Loughran et al. 2004).

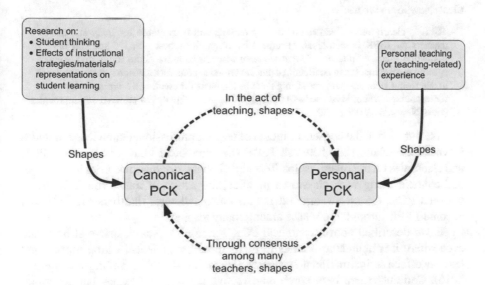

Fig. 3.1 Personal-canonical PCK synergy model

We have tried to combine personal PCK from many teachers in many parts of the USA, looking for aspects that transcend context.

It is important to note, however, that canonical PCK is *not* dependent on personal PCK. It can accumulate entirely from the types of research listed in the model. Unfortunately, the canon has large gaps—that is, some topics lack widely accepted PCK, and even those that do often focus more on student thinking than effective instructional strategies. For example, canonical PCK about students' thinking related to force and motion is abundant (e.g., Gunstone and Watts (1985) synthesized several studies more than 30 years ago), but our literature search (Smith and Plumley 2016) suggests that knowledge of effective ways to teach elementary students about the particle model of matter is scarce.

A second aspect of synergy is represented in the assertion that personal PCK can be shaped by canonical PCK. That is, a teacher may read about a particular aspect of student thinking or about a particularly effective instructional strategy. Then, in the act of teaching (or preparing to teach), that knowledge may be transformed by a teacher's experience. Personal PCK is not, however, dependent on canonical PCK. This phenomenon is particularly evident at the university level, where professors have abundant disciplinary *content* knowledge (e.g., physics or biology knowledge) but, historically, have had little or no exposure to canonical PCK. However, they still have personal PCK by virtue of their teaching experience. Through their teaching alone, they form ideas about effective instructional strategies and patterns in student thinking—that is, they form personal PCK.

The 2012 PCK Summit (Carlson et al. 2015) generated a consensus model of teacher professional knowledge that reflects the notion of synergy. Although canonical PCK is not included by name, a component named "topic-specific professional knowledge" (or TSPK) is the same construct. Describing the model, Gess-Newsome wrote:

> TSPK is clearly recognized as codified by experts and is available for study and use by teachers…. TSPK is canonical, generated by research or best practice and can have a normative function in terms of what we want teachers to know about topic- and context-specific instruction. It can be identified and described to construct measures, tests, or rubrics to determine what teachers know, might act as the basis for creating a learning progression for teachers, and should be used as a framework for the design of professional development. (Gess-Newsome 2015, p. 33)

Like our model, the consensus model of teacher professional knowledge includes a synergistic relationship between TSPK (i.e., canonical PCK) and personal PCK and elaborates on the relationships. Teacher's beliefs, orientations, prior knowledge, and context, along with classroom practice, are active in transforming TSPK to personal PCK. Gess-Newsome (2015) also acknowledges that personal PCK can become TSPK through consensus among many teachers.

As we described above, canonical PCK for some topics is sparse at best, and even where it is abundant, it tends to focus more on student-thinking aspects and less on effective instructional strategies (Hayes et al. 2017; Smith and Plumley 2016). Certainly there is research on effective teaching strategies, but not topic-specific teaching strategies. If one imagines a matrix with NGSS disciplinary core ideas as rows and research on student thinking and instructional strategies as columns, we suspect the majority of cells would be empty or lightly populated.

We hypothesized that personal PCK, collected and codified from many teachers, could fill gaps in canonical PCK. Over 3 years, we tested that hypothesis in the context of the two NGSS disciplinary core ideas mentioned in the Introduction. First, we reviewed empirical research studies and practitioner-oriented literature and then synthesized information on student-thinking and topic-specific instructional strategies, generating canonical PCK (e.g., Smith et al. 2017). Next, we surveyed and interviewed teachers to collect their PCK, exploring whether common elements in their knowledge might rise to the level of canonical PCK as we have defined it and whether it would fill gaps in a topic-specific canon. We were not looking for evidence of canonical PCK in teachers' responses, but rather testing whether we could generate canonical PCK from their combined responses. The work has yielded important insights. Before discussing these, it is necessary to briefly describe the methods used to elicit personal PCK from teachers.

3.3 Eliciting Personal PCK from Teachers

Attempts to elicit PCK from teachers, which are almost as old as the construct itself, face a common obstacle. Shulman wrote in one of his earliest papers conceptualizing PCK: "Practitioners simply know a great deal that they have never even tried to articulate" (Shulman 1987, p. 12). His statement is true 30 years later. Teachers seldom need to articulate their PCK for themselves, and they are rarely, if ever, asked to articulate it for others. Consequently, their PCK tends to be tacit (Cohen and Yarden 2009; Henze and Van Driel 2015; Loughran et al. 2004, 2008). Despite this formidable obstacle, testing our synergy hypothesis required eliciting and characterizing teachers' personal PCK. Like other PCK researchers, we found affordances and limitations in a survey approach. Survey questions were based on the CoRe methodology (Loughran et al. 2004) used widely in studies of teacher knowledge (e.g., Alvarado et al. 2015; Williams and Lockley 2012).

We administered a web-based survey to grades 4–6 teachers from several states about their topic-specific PCK related to the small particle model of matter and interdependent relationships in ecosystems. Some questions asked about teachers' knowledge of student thinking, for example, "Please describe the ideas or misconceptions your students have that make it difficult for them to learn about **the particle model of matter**."[3] Others asked about their instructional strategies, for example, "Please describe a question or activity you use to find out what ideas students already have about the interdependent relationships in ecosystems **before** you

---

[3] Teachers were presented with this question only if they had already responded that students do have misconceptions that make it difficult for them to learn about the small particle model or interdependent relationships in ecosystems.

begin teaching about it."[4] The survey allowed respondents to upload documents they used in their teaching, including laboratory activities and worksheets. Respondents were also encouraged to share other resources, for example, online simulations and videos.

The survey forced teachers to compartmentalize their knowledge (e.g., they responded to separate survey questions about student-thinking patterns and instructional activities). From an analysis standpoint, this feature was an affordance. However, survey responses did not represent how different types of personal PCK related to each other. As illustrated by the examples above, the survey asked respondents to describe misconceptions and instructional activities separately, rather than explain which activities they use to address their students' misconceptions. In addition, responses tended to be vague, lacking detail needed to characterize a teacher's PCK adequately.

We ultimately found a combined survey-and-interview approach most effective. Teachers first completed the web-based survey. We then conducted a follow-up telephone interviews with survey respondents, during which we probed on each of their survey responses.[5] Before the interview, each interviewee received his or her survey responses by email and was encouraged to have them on hand during the interview. The interview followed essentially the same structure as the survey, but researchers probed for elaboration of survey responses that were unclear and for connections among compartmentalized responses. For example, a survey respondent may have written "I ask questions" when describing a particular activity. During the interview, a researcher prompted the respondent to name the specific questions and asked for typical student responses. Similarly, a survey respondent who provided only a sentence or two about an activity was asked to expand upon their description during the interview. Interviewers also asked respondents how they used the resources that they had uploaded in their survey responses (e.g., lesson plans, student handouts). Table 3.1 shows an example of survey and follow-up interview responses from one teacher.

Researchers merged the survey responses and interview transcripts into a single document for each individual, with the response for each survey question arranged next to relevant parts of the interview transcript. Researchers coded the documents using a two-dimensional framework: (1) PCK type (e.g., student thinking, instructional activity) and science content (i.e., the discrete science ideas in each concept[6]). Together, researchers first coded several documents to establish common understanding of the framework and then divided and coded the rest of the documents as individuals using qualitative analysis software (Dedoose). After all documents were coded, two researchers—one for each topic—analyzed the documents for

---

[4] Teachers were presented with this question only if they had already responded that they try to elicit student thinking before instruction begins.

[5] The findings in this chapter are based on 42 combined survey-interviews, about equally split between the two topics (small particle model of matter and interdependent relationships in ecosystems).

[6] The science ideas are described in Smith and Plumley (2016) and Hayes et al. (2017).

Table 3.1 Survey and follow-up interview responses

Survey	Interview
Please describe a question or activity you use to find out what ideas students already have about the particle model of matter before you begin teaching about it I just ask them what a gas is, a solid is, and a liquid is, and we brainstorm properties that each have	*Interviewer (I)*: You said you do a bit of a brainstorming about the different states and properties. What kinds of things in that brainstorming session do you hope to accomplish? *Teacher (T)*: Well, we put something on the whiteboard, and we talk about, "Okay, give me examples of solids. Give me examples of liquids. Give me examples of gases." Then we put all those down. Then we discuss how they are similar, how they are different, and we kind of then come up with the fact that—I'll say to them, "But did you know that all of these have something in common because they're all made up of matter that is composed of these particles." We kind of just go through that thing. I'm not really teaching it to them at this point, just letting them know that they all do have something in common *I*: When the students are giving you examples, what are some of the typical responses that they come up with? What kinds of things do they say about the properties? *T*: Well, they'll say, for example, with the solids, "Well, solids are things that you can touch." Then they'll break it down into things like, "Well, things that are wood or things that are metal and that sort of thing." Liquids, they'll break those down into just common household things—soda, water, milk, all of those. They're really not getting to the actual properties themselves, but they're listing everything that they can that has to do with the solids and the liquids—household items that they know of.

frequency of PCK and summarized the findings. The research team reviewed the summaries to confirm that findings aligned with their own experience interviewing teachers and coding the combined survey-interview response documents.

3.4 Unrealized Synergy Potential

As evidenced by the preceding discussion, some aspects of our work have been challenging. However, we began investigating the synergy hypothesis optimistically. We had already identified large gaps in canonical PCK before we surveyed and interviewed teachers, and the potential to fill those gaps with personal PCK from many teachers—that is, the path in our model from personal PCK to canonical PCK—was appealing. However, our work thus far does not support the hypothesis. When responding to survey and interview questions, teachers tend to report instructional approaches similar to those in the literature—variations rather than new, effective ways of teaching concepts. Regarding PCK about student thinking, teachers tend not to describe student misconceptions reported in literature, and they generally do not identify new misconceptions. In short, we claim that the hypothesized potential for synergy between canonical and personal PCK is not being realized. In the discussion that follows, we support this claim with evidence from our studies. We address PCK in two broad categories—instructional PCK first, followed by knowledge about student thinking.

3.4.1 Instructional PCK

Before the NGSS, the small particle model of matter (SPM) and the phenomena it explains were taught mainly in middle grades and higher (American Association for the Advancement of Science 1993; National Research Council 1996). Not surprisingly, the literature on instructional activities for SPM in the elementary grades is very limited. In contrast, an abundance of instructional activities exists for interdependent relationships in ecosystems. Looking across many of these activities in the empirical and practitioner literature, broader instructional approaches emerged, including engaging students with a scenario of an ecological disturbance, constructing and analyzing foods webs, and examining a particular ecosystem. Generally, instructional activities reported by teachers appeared to be variations on one of those broader approaches, rather than novel ones. We illustrate this pattern with several examples below.

The scenario approach appeared in numerous forms in the literature, including roleplay simulations, thought experiments, online simulations, and videos. Roleplay simulations, in which students take on organism roles and act out a scenario, were particularly prevalent. However, when describing seemingly similar activities, authors and teachers differed in their procedures and intended student learning outcomes. Table 3.2 includes examples of how the widely used "Oh Deer!" roleplay activity was referred to in the practitioner literature and by a teacher. In the example from the literature, predation is not introduced in the activity itself, whereas the teacher described this addition, with some students playing the part of wolves. Acknowledging the prevalence of the activity, but not the differences among versions, another teacher said:

> I know ["Oh Deer!"] shows up in a lot of different curriculums under different names. I just attached the one that I could find online to make it easier.

Interestingly, the version that the teacher attached to her survey response differed from the implementation she described, in that the attached version did not introduce predation (see Table 3.2). The larger point is that teachers reported minor variations of the "Oh Deer!" scenario, for example, rather than wholly different activities.

Creating a food web, or in some cases a food chain, appears frequently in the literature as an instructional approach to represent and examine trophic relationships. Likewise, teachers often describe individual students drawing a web, arranging cards labeled with organisms, or using an online interactive simulation. One oft-cited whole class activity uses string, or yarn, to trace connections among organisms (typically with individual students playing the role of a population of organisms—e.g., rabbits) and simulate the effects of disturbances to an ecosystem (Appel et al. 1982; Camp 1995; Clement et al. 1997; Kuhn 1971). Table 3.3 includes two summaries of the string web activity from the practitioner literature, as well as two teachers' accounts of how the activity takes shape in their classroom. Again, the teacher accounts include variations but not new approaches.

Table 3.2 Example uses of the "Oh Deer!" instructional activity

Source	"Oh Deer!" implementation description
Rockow (2007, p. 19)	*Students play "Oh Deer!" (Dalton, 1992), an interactive game that demonstrates what happens to a population of animals when there are more animals than the ecosystem can support…. Students sometimes have the misconception that if the population of a species declines, extinction will follow. This activity shows students that populations can decline and then rebound, without leading to extinction. At the end of this period I talk about the limitations of this activity as a model for an ecosystem…*
Teacher A	*Briefly, it's basically, a fourth of your students become "deer," and the other three-fourths become the "resources." You can be "food," "water," and "shelter." There's an accompanying gesture that you make for the deer to know what you are. The deer can't see what's out there until they turn around after you say, "Oh Deer." Then they run across the field. They're holding whatever sign it is, and trying to find a matching sign on the other side…. What we'll do to create some interesting years in our data chart will be, this year there's a "drought," and no one can be "water." This year there's a fire, so no one can be "shelter." Any deer looking for those things would perish, and we would be able to be like, "Oh, that's the year of the fire!"…. Just have 'em thinking about some other factors that might influence the population. Eventually, to make a little more game out of it, we'll introduce a predator. We usually call it a "wolf." … It's trying to find one deer per year to bring back to its den. Its population begins to shift as well…. Usually, what happens, is the wolf, somehow, they're able to draw the resources, or whatever; but, usually, the game ends by the wolf population just dominating the deer population. The deer die out. The following year, the wolves die out. We talk about, "How would that look in nature? Would that really happen if we have 20 wolves and two deer, and the deer was the only food source?" I think that's one of the problems with the whole activity is that most organisms have many different food sources…. In some form or another, I feel like I've done "ecosystems" for 8 or 9 years. I've always done "Oh Deer." It hits so many checkboxes where it's very focused on the learning objectives*

Examining a particular ecosystem—that is, observing, or researching an ecosystem and its components, and situating instructional activities in that ecosystem—emerged as another common instructional approach in the literature. One example is a comprehensive unit guide designed by the Long Island Pine Barrens Society (1998) to integrate classroom investigations with outdoor experiences related to the Long Island Pine Barrens. This guide includes the string food web simulation described previously, in which all organisms are native to this particular ecosystem (e.g., Pitch Pine, Tiger Beetle, Red Fox). Another example is the Smithsonian Institute's *Art to Zoo* publication (1996), which features instructional resources designed to contrast the coral reef of the Caribbean and the rocky coast of Maine. Within these activities, students examine trophic relationships and consider the impacts of both biotic and abiotic factors.

Similarly, teachers described concentrating a portion of their instruction on a particular, often local, ecosystem. Some teachers reported focusing on a local ecosystem as a starting point for instruction, because students have familiarity through prior experience, or the class could take a field trip to provide an experience upon which students can draw. One teacher described what influenced her decision

Table 3.3 Example uses of the string food web instructional activity

Source	String food web implementation description
Clement et al. (1997, p. 28)	*Using a large ball of yarn, start with water and sunlight and ask what members of the wetland use these things. Connect students with yarn as they demonstrate relationships. Cut the yarn whenever it becomes cumbersome. Eventually it should be clear that all members of the ecosystem are connected. Try tugging on one link of the web and seeing how many students can feel it. If each student who feels the tug pulls on the lines he or she is holding, the original tug will ripple through the whole community just as wetland disturbances affect many organisms*
Kuhn (1971, p. 832)	*As the data is analyzed, each student assumes the role of an organism of the community—one might be a green alga, another a water flea, another a catfish, another a snail, etc. As each relationship is established, a line is strung, e.g., between the "producer" organism and a primary consumer. Other relationships can be established in a similar manner. As the analysis continues, the existing relationships become evident; one primary consumer may feed upon several producers; a third-order consumer may feed upon several other animals. The complexity of the food web becomes strikingly evident, and the visual impact is substantial*
Teacher B	*Everybody, all the kids, all have a role. I pass out several different roles, or if I don't—like if I have a huge class, I pair them up… Everybody has a piece of the string. We pass it out, and I weave it through the kids, basically by saying, "Okay. You're a—," I'm just making up things right now—but, "You're a daisy." Okay. Well, a daisy is eaten by the rabbit. I'll throw the string across to the rabbit. The rabbit has to get it. Okay. Well, now the rabbit's gonna get eaten by a couple of different things*
	By the time we're done, it's a web. It is truly. It looks like a web the way the string is passed around. The kids see the interconnection that way. Then what's also neat is you can see the web collapse when an animal misses, 'cause all of a sudden I'll say, "Okay. Somebody sprayed [herbicide] and all the plants died. If you are a plant or you are a producer, please drop." They'll physically drop the string. Well, then they see the whole thing go. A lot of times then I'll say, "Okay. What happens? How does this affect these animals? How does this affect these animals? Okay. How is this gonna affect an herbivore?" Well, they're not gonna have anything to eat. Okay. "Therefore?" "Well, they'll die." "Okay. Then when they die, what's gonna happen?" "Well, a decomposer will take over, but we won't have enough decomposers to take care of things."
	It's funny because when we really look at it—after they're done with the physical activity, we go back, and we answer some questions in our notebook about it. I'll make them think about if all the producers disappear, what's gonna happen to an herbivore? What's gonna happen to the land itself? What's gonna happen to—we take it apart as to each role and what happens
Teacher C	*There's one, also, with food webs that we've tried, where you give them yarn, and they go from place to place. It ends up kind of a crazy web. Naturally, it doesn't really work perfectly. It's not a true food web, but they can see how complicated the food web can truly be. That's more the point of it is that those webs are pretty complicated…. You assign each child, "You're this kinda plant; you're this kinda bird," everybody has a role. Then the bird could fly from a plant, over to a different plant. Then the cat can come, and he has to stand with the bird, and then the bird is eaten, so the bird is still. It won't move anymore, and then something comes and gets the cat. As they move from place to place and you clear out the classroom, and they have a long piece of yarn—when the cat goes to the bird, the bird would hold on to the yarn, that part of the yarn, and then the cat could move on, but it took that to the bird. It's a three-dimensional kind of thing, and as I said, it doesn't work perfectly, but it does allow them to see the complexity*

to have students identify abiotic and biotic factors in a deciduous forest at the beginning of her ecosystems unit:

> It's mainly the fact that we live in Ohio ... [Deciduous forests are] a very common ecosystem. We have this property that we visit that is a deciduous forest. They've all had that concrete experience of being there, which I find is really important for them to engage in the conversation.

3.4.2 Student-Thinking PCK

The vast majority of topic-specific PCK research literature focuses on student thinking. In these studies, researchers typically presented students with a phenomenon and asked them to explain it (e.g., Abraham et al. 1994; Helldén 1998; Leach et al. 1992). Some studies involved students responding to multiple-choice assessments, with common misconceptions as answer choices; others had students write (or draw) questionnaire responses or provide oral explanations in an interview. Whatever the format, in these studies, researchers presented students with something to explain, determined students' misconceptions, and described them in their reports. Excerpts from these types of studies, including the elicitation tasks and examples of resulting misconceptions, are in Table 3.4.

It appears that teachers rarely do these types of elicitation activities with students to find out what they think at the beginning of instruction. Instead, when teachers elicit student ideas, they tend to be about a broad topic (e.g., matter) using methods such as a KWL chart (What do I **K**now?, What do I **W**ant to know?, What did I **L**earn?). The box below includes an example of a teacher describing the use of a KWL.

KWL Description

Interviewer: Going on to the next idea, could you lead me through how you set up this KWL chart for the concept of matter? You said it's pretty informal. Could you just give me a little better idea of how that looks in the class?

Teacher: Yeah. A lot of times, I will just make an anchor chart, so a piece of chart paper, and put a KWL on there, and we just divide it. We talk about, "Well, if I say 'matter,' what do you know about matter? What can you tell me about matter?" Sometimes—it depends if those teachers under us actually taught their science or not. Sometimes they have things that we already know.

Interviewer: Okay. What would you say your students do already know?

Teacher: This year, they didn't know a lot... It really varies from year to year. Sometimes they know that—sometimes they know, "Hey, I know that there's matter out there." I don't think I've ever had anybody come in being able to tell me that matter is particles that are too small to be seen, and sometimes really large, and it's everywhere, and it has mass, and it takes up space. No one's ever been able to really tell me the whole thing. There's always something to learn.

Table 3.4 Elicitation tasks and misconceptions from literature

Source	Elicitation task	Resulting problematic student thinking
Abraham et al. (1994)	A cube of sugar is added to a glass of water. The cube of sugar dissolves. Describe what happens to the sugar cube as it dissolves. Picture 1 shows how the glass looked after the sugar cube was added to the water. In glass 2 draw how the sugar cube and water would look after half of the sugar dissolved. In glass 3 draw how the glass of water would look after all the sugar dissolved. Water — Sugar Cube — 1 2 3	When sugar dissolves, the water in the glass absorbs the sugar. When the sugar dissolves, it undergoes a phase change, melts, or evaporates.
Helldén (1998)	What do you think will happen to the plant in the box if we plant it there and glue the lid on?	Students expected the plant to die immediately in the sealed, transparent box. Students viewed the plant as the "end station" for matter, describing how the necessary resources for growth and survival would be consumed but not attending to what plants produce.
Leach et al. (1992)	One summer there is a drought. A lot of the grass and crops die. What do you think might happen as result of this? Explain as carefully as you can.	Students often appeared to consider effects on individual organisms, rather than on a population. Students appeared more likely to trace effects up through the food web than down, demonstrating differences in their reasoning based on the trophic level being affected.

KWL charts can play an effective role in instruction, but as elicitations, they provide the teacher with little useful information about students' topic-specific misconceptions. The student thinking that typically arises in response to a KWL chart about a broad topic is something we refer to as a "missing conception"; that is, an idea that students are not familiar with rather than a *misconception* about the idea. Similarly, when asked about topic-specific student misconceptions, teachers instead typically respond that students "don't know about that," with "that" being either the concept itself or related vocabulary. Table 3.5 includes several examples of missing conceptions expressed by teachers.

Similarly, when attempting to describe misconceptions, teachers often report that their students have difficulty understanding the topics of SPM and interdependence because phenomena involve components and changes too small or too slow to see.

Table 3.5 Missing conceptions about interdependent relationships and SPM

Source	Missing conception
Teacher D	*Again, my students do not have knowledge of the natural world around them. They have the idea that animals eat what you feed them—basically a pet mentality.*
Teacher E	*They haven't been exposed to [decomposition] before. They haven't really thought about the roles of, say, the fungus with the mushrooms or the earthworms. They haven't really considered those roles before, and so they don't understand that they break things down.*
Teacher F	*Like I said, they might use it in a different terminology, where they have no idea that matter means everything that's around us, everything that we're seeing. There's different types of matter that make up the universe. They're using it in, "Well, I didn't do this, so why does it matter?" I'm like, "No, that's not the type of matter I'm talking about." They're like, "What do you mean?" I'm like, "Matter—everything that makes up stuff."*

Table 3.6 Developmental challenges for interdependent relationships and SPM

Source	Developmental challenge
Teacher G	*Students often do not have any concrete experience with decomposition, other than molding food which is promptly thrown away. They can't see many of the microorganisms and often don't get to see the process from start to finish, which leads to misconceptions and just incorrect ideas.*
Teacher H	*I think it's hard for kids to understand anything that doesn't happen immediately. It's something that I also see, for example, with growing plants and things like that, things that take time. In their world, a long time is a week.*
Teacher I	*Sometimes if they can't see the particles, or anything for that matter, then it's difficult for some to grasp that there is actually stuff they can't see.*

We call this type of student thinking a "developmental challenge" rather than a misconception, because it relates to something that is difficult for students of this age to grasp, not an incorrect idea. Examples of developmental challenges shared by teachers are in Table 3.6.

In the work discussed in this chapter, we asked teachers about student thinking related to two specific topics. The teachers did sometimes provide misconceptions (in the sense that we use the term) that addressed concepts *prerequisite* to the topics. For example, in SPM we were interested in students' misconceptions about the particle model and about the phenomena that the particle model helps explain. However, teachers frequently described misconceptions related to states of matter instead, such as thinking that gases are not matter because they cannot be touched like liquids and solids. Similarly, for interdependent relationships in ecosystems, teachers spoke of students' misconceptions about components of the ecosystem, rather than how they interact. For example, many teachers described that their students think that they, themselves, are producers because they can make (i.e., cook) their own food. This finding—that teachers are more familiar with misconceptions at the boundaries of the key ideas about a topic—may be specific to the ideas we are investigating. More research is required in other topics. To repeat our general finding, however, teachers typically do not report topic-specific student misconceptions outside of those in the literature. Consequently, their potential to fill gaps in canonical PCK about student thinking appears limited.

3.5   Discussion

In concluding this chapter, we first acknowledge that canonical PCK is not universally accepted. We attribute the resistance in part to semantics; the term "canonical" evokes strong negative reactions in some. The idea that PCK exists outside of an individual is unacceptable to others. But we suspect that most would agree our field has identified prominent, enduring patterns in students' topic-specific thinking. Young students tend to think that moving objects always stop, that dissolving solids cease to exist, and that decaying matter just "goes away" without any biochemical action. Similarly, the field has identified effective instructional strategies for helping students reconcile these ideas with accepted scientific concepts. When students slide a block across progressively smoother surfaces, the experience (if well facilitated by a knowledgeable teacher) can challenge their idea that moving objects always stop. Weighing the mass of a solid and liquid before and after dissolving suggests that the solid does not cease to exist. And careful observations of composting can open students' minds to the possibility of invisible processes they had never considered. These are instances of what we call canonical PCK, but the term is not as important as what it represents—widely accepted knowledge about how students think and learn about a topic.

In exploring the synergy hypothesis, we focused on the potential for accumulated personal PCK to fill gaps in canonical PCK. We gave little attention to the other aspect of synergy—that canonical PCK can become personal as one takes it up and uses it in teaching. However, we see little evidence of this aspect of synergy either. Reports from teachers suggest minimal exposure to empirical literature on student thinking, which is not surprising. Apart from the Driver books mentioned earlier, attempts to synthesize the literature are infrequent, and attempts to make the knowledge accessible to teachers are even less frequent. Regarding instructional PCK, we described how teachers report using some of the same approaches we found in practitioner literature, but we saw little evidence that teachers got them *from* the literature. Rather, teachers tended to reference the results of Internet searches, citing sources such as BrainPOP, StudyJams, and Teachers Pay Teachers, among others.

We are unable to support either relationship in our synergy hypothesis, but a caveat is in order. We synthesized canonical PCK by collecting, reviewing, and summarizing empirical and practitioner literature. These tasks were straightforward, not substantially different from any other literature review. Eliciting personal PCK from teachers was far more challenging. We found the combined survey-and-interview approach promising, but we are uncertain that it adequately addresses the tacit nature of teachers' PCK documented elsewhere (e.g., Cohen and Yarden 2009;

Henze and Van Driel 2015; Loughran et al. 2004, 2008). Other means of eliciting personal PCK may yield different findings, perhaps even evidence for the synergy hypothesis.

We did find complementarity between empirical and practitioner literature, at least for the interdependence topic.[7] Empirical literature focused heavily on student thinking, practitioner literature almost exclusively on instructional activities. Together, they form a richer knowledge base for teaching the topic than either does alone. What is lacking are efforts to (1) synthesize within and across these types of literature, and (2) make the product accessible to teachers in a form they are likely to use. Also lacking in either type of literature is knowledge about the affordances and limitations of an activity in terms of student thinking. Why is this activity important? What can it accomplish for students? What should come before and after it?

For all of these reasons, we are encouraged by the vision of educative curriculum materials—that is, curriculum materials that incorporate features designed to support teachers' purposeful enactment of the materials (Davis et al. 2014; Davis and Krajcik 2005). Educative features can make canonical PCK available to teachers when and where they need it. An example is narrative text for the teacher that explains how a unit sequence develops student understanding across activities. Within an activity, these features can highlight the activity's function in the broader unit, making it more likely a teacher will capitalize on the activity's affordances and compensate for its limitations. Another type of educative support can point teachers to patterns of student thinking the teachers can leverage and others the teacher should be prepared to challenge appropriately. In short, educative curriculum materials can do the work that teachers do not have time, and perhaps the background, to do themselves. In this way, educative features can catalyze the pathway from canonical PCK to personal PCK.

The work described in this chapter is ongoing, the findings tentative. Our goal was to put forth the synergy hypothesis and describe our findings to date in broad strokes. Given the large gaps in canonical PCK within and across topics, the synergy hypothesis is still appealing. An underdeveloped canon limits the work of teachers, teacher educators, and curriculum developers. For example, educative curriculum materials can make canonical PCK available to teachers only if it already exists. Topics with a weak canon require foundational research before educative curriculum materials can play their proper role. The question is whether such research can focus on accumulating topic-specific personal PCK, as we have attempted, or whether it must include student-level studies. The former is almost certainly more expedient and less costly, but our work suggests the latter may be necessary. The field needs more research on both fronts.

---

[7] We discussed earlier in the chapter that the small particle model has not been taught widely in elementary grades prior to the NGSS. Consequently, practitioner literature for this topic in these grades is lacking.

Acknowledgment   This research was supported by the National Science Foundation Grant DRL-1417838. Any opinions, findings, and conclusions or recommendations expressed in this paper are those of the authors and do not necessarily reflect the views of the National Science Foundation.

References

Abraham, M. R., Williamson, V. M., & Westbrook, S. L. (1994). A cross-age study of the understanding of five chemistry concepts. *Journal of Research in Science Teaching, 31*(2), 147–165. https://doi.org/10.1002/tea.3660310206.

Alonzo, A. C., & Kim, J. (2016). Declarative and dynamic pedagogical content knowledge as elicited through two video-based interview methods. *Journal of Research in Science Teaching, 53*(8), 1259–1286. https://doi.org/10.1002/tea.21271.

Alvarado, C., Cañada, F., Garritz, A., & Mellado, V. (2015). Canonical pedagogical content knowledge by CoRes for teaching acid–base chemistry at high school. *Chemistry Education Research and Practice, 16*(3), 603–618. https://doi.org/10.1039/C4RP00125G.

American Association for the Advancement of Science. (1993). *Benchmarks for science literacy: Project 2061*. New York, NY/Oxford, UK: Oxford University Press.

Appel, G., Jaffe, R., Cadoux, M., & Murray, K. (1982). *The growing classroom: A garden-based science and nutrition curriculum for 2nd through 6th grades. Book 2: Science*. Retrieved from http://eric.ed.gov/?q=ecology+interdependence&ff1=subElementary+School+Science&id =ED239918

Camp, C. A. (1995). *Invitations to interdependence: Caught in the web. Teacher-friendly science activities with reproducible handouts in English and Spanish. Grades 3–5. Living Things Science Series*. South Deerfield, MA: Ash Grove Press, Inc. Retrieved from http://eric.ed.gov/ ?q=Ecosystems+elementary+students&pg=13&id=ED392641

Carlson, J., Stokes, L., Helms, J., Gess-Newsome, J., & Garder, A. (2015). The PCK Summit: A process and structure for challenging current ideas, provoking future work, and considering new directions. In *Re-examining pedagogical content knowledge in science education*. New York, NY: Routledge.

Clement, J., Sigford, A., Drummond, R., & Novy, N. (1997). *World of fresh water: A resource for studying issues of freshwater research*. Retrieved from http://eric.ed.gov/?q=ecosystems+unde rstanding&ff1=subScience+Instruction&pg=4&id=ED479305

Cohen, R., & Yarden, A. (2009). Experienced junior-high-school teachers' PCK in light of a curriculum change: "The cell is to be studied longitudinally". *Research in Science Education, 39*(1), 131–155. https://doi.org/10.1007/s11165-008-9088-7.

Dalton, P. (1992). *ProjectWILD*. Bethesda, MD: Western Regional Environmental Education Council, Inc..

Davis, E. A., & Krajcik, J. S. (2005). Designing educative curriculum materials to promote teacher learning. *Educational Researcher, 34*(3), 3–14.

Davis, E. A., Palincsar, A. S., Arias, A. M., Bismack, A. S., Marulis, L. M., & Iwashyna, S. K. (2014). Designing educative curriculum materials: A theoretically and empirically driven process. *Harvard Educational Review, 84*(1), 24–52.

Driver, R. (1994). *Making sense of secondary science: Research into children's ideas*. London, UK/New York, NY: Routledge.

Driver, R., & Easley, J. (1978). *Pupils and paradigms: A review of literature related to concept development in adolescent science students*. Retrieved from http://www.tandfonline.com/doi/ pdf/10.1080/03057267808559857

Driver, R., Guesne, E., & Tiberghien, A. (Eds.). (1985). *Children's ideas in science*. Milton Keynes, UK/Philadelphia, PA: Open University Press.

Gess-Newsome, J. (2015). A model of teacher professional knowledge and skill including PCK: Results of the thinking from the PCK Summit. In A. Berry, J. Loughran, & P. J. Friedrichsen (Eds.), *Re-examining pedagogical content knowledge in science education*. London, UK: Routledge.

Gunstone, R., & Watts, M. (1985). Force and motion. In R. Driver, E. Guesne, & A. Tiberghien (Eds.), *Children's ideas in science* (pp. 85–104). Milton Keynes, UK/Philadelphia, PA: Open University Press.

Hayes, M. L., Plumley, C. L., Smith, P. S., & Esch, R. K. (2017). *A review of the research literature on teaching about interdependent relationships in ecosystems to elementary students*. Chapel Hill, NC: Horizon Research, Inc..

Helldén, G. F. (1998). *A longitudinal study of students' conceptualization of ecological processes*. Retrieved from http://eric.ed.gov/?q=ecology+%22student+thinking%22&pg=2&id=ED440882

Henze, I., & Van Driel, J. H. (2015). Toward a more comprehensive way to capture PCK in its complexity. In A. Berry, J. Loughran, & P. J. Friedrichsen (Eds.), *Re-examining pedagogical content knowledge in science education*. London, UK: Routledge.

Kuhn, D. J. (1971). Simulation of a food web. *School Science and Mathematics, 71*(9), 831–833.

Leach, J., Driver, R., Scott, P., & Wood-Robinson, C. (1992). *Progression in understanding of ecological concepts by pupils aged 5 to 16*. University of Leeds, Centre for Studies in Science Education. Retrieved from http://www.opengrey.eu/item/display/10068/481018

Lee, O., Eichinger, D. C., Anderson, C. W., Berkheimer, G. D., & Blakeslee, T. D. (1993). Changing middle school students' conceptions of matter and molecules. *Journal of Research in Science Teaching, 30*(3), 249–270.

Long Island Pine Barrens Society. (1998). *The Long Island Pine Barrens: A curriculum & resource guide*. Retrieved from http://eric.ed.gov/?q=ecosystem+lesson+plans&ff1=subEcology&id=ED436344

Loughran, J., Mulhall, P., & Berry, A. (2004). In search of pedagogical content knowledge in science: Developing ways of articulating and documenting professional practice. *Journal of Research in Science Teaching, 41*(4), 370–391.

Loughran, J., Mulhall, P., & Berry, A. (2008). Exploring pedagogical content knowledge in science teacher education. *International Journal of Science Education, 30*(10), 1301–1320.

National Research Council. (1996). *National science education standards: Observe, interact, change, learn*. Washington, DC: National Academy Press.

NGSS Lead States. (2013). *Next generation science standards: For states, by states*. Washington, DC: National Academies Press.

Osborne, R. J., & Cosgrove, M. M. (1983). Children's conceptions of the changes of state of water. *Journal of Research in Science Teaching, 20*(9), 825–838. https://doi.org/10.1002/tea.3660200905.

Rockow, M. (2007). Tabizi pythons and clendro hawks: Using imaginary animals to achieve real knowledge about ecosystems. *Science Scope, 30*(5), 16–22.

Russell, T., Harlen, W., & Watt, D. (1989). Children's ideas about evaporation. *International Journal of Science Education, 11*(5), 566–576. https://doi.org/10.1080/0950069890110508.

Shulman, L. (1987). Knowledge and teaching: Foundations of the new reform. *Harvard Educational Review, 57*(1), 1–23.

Smith, P. S., & Plumley, C. L. (2016). *A review of the research literature on teaching about the small particle model of matter to elementary students*. Chapel Hill, NC: Horizon Research, Inc..

Smith, P. S., Plumley, C. L., & Hayes, M. L. (2017). Much ado about nothing: How children think about the small-particle model of matter. *Science and Children, 54*(8), 74–80.

Smithsonian Institution. (1996). Contrasts in blue: Life on the Caribbean coral reef and the rocky coast of Maine. *Art to Zoo: Teaching with the Power of Objects*. Retrieved from http://eric.ed.gov/?q=Elementary+lesson+ecosystem&pg=3&id=ED409253

Tytler, D. R., & Peterson, S. (2000). Deconstructing learning in science—Young children's responses to a classroom sequence on evaporation. *Research in Science Education, 30*(4), 339–355. https://doi.org/10.1007/BF02461555.

Veal, W. R., & MaKinster, J. G. (1999). Pedagogical content knowledge taxonomies. *Electronic Journal of Science Education, 3*(4). http://ejse.southwestern.edu/article/view/7615.

Williams, J., & Lockley, J. (2012). Using CoRes to develop the pedagogical content knowledge (PCK) of early career science and technology teachers. *Journal of Technology Education, 24*(1), 34–53.

Chapter 4
From Budgets to Bus Schedules: Contextual Barriers and Supports for Science Instruction in Elementary Schools

Cathy Ringstaff and Judith Haymore Sandholtz

Abstract Improvements in teachers' pedagogical content knowledge are critical in improving science education but may be insufficient to support and sustain instructional changes. This chapter describes how contextual factors influenced teachers' use of research-based instructional strategies learned in professional development. The research draws on survey, observational, and interview data collected from 135 teachers who participated in four different intensive professional development programs that were situated in small, rural school districts with high-need student populations and that extended over 3 years. Each program had a slightly different STEM focus, but all four programs aimed to improve teachers' pedagogical content knowledge in science and to foster their use of research-based instructional strategies in science. Across programs, teachers' science content knowledge, pedagogical content knowledge, and self-efficacy increased over the course of the professional development. Overall, teachers more frequently used research-based practices for teaching science. But contextual factors varied substantially across schools and districts and both fostered and hindered teachers' science instruction. The most influential contextual factors included time for planning and collaboration, time for science instruction, administrator support, access to resources, and regional constraints. Identifying the contextual factors that influence teachers' use of pedagogical content knowledge gained through professional development is the first step for formulating strategies for supporting and sustaining teacher change.

Keywords PCK · Science education · Teacher professional development · Contextual factors

C. Ringstaff (✉)
WestEd, San Francisco, CA, USA

J. H. Sandholtz
University of California, Irvine, CA, USA

© Springer International Publishing AG, part of Springer Nature 2018
S. M. Uzzo et al. (eds.), *Pedagogical Content Knowledge in STEM*,
Advances in STEM Education, https://doi.org/10.1007/978-3-319-97475-0_4

4.1   Introduction

Improvements in teachers' pedagogical content knowledge are critical in improving science education but may be insufficient to support and sustain instructional changes, particularly in rural schools. This chapter describes how contextual factors influenced teachers' use of research-based instructional strategies learned in professional development. The research draws on survey, observational, and interview data collected from 135 teachers who participated in four different intensive professional development programs that were situated in small, rural school districts with high-need student populations and that extended over 3 years. Each program had a slightly different STEM focus, but all four programs aimed to improve teachers' pedagogical content knowledge in science and to foster their use of research-based instructional strategies in science. Identifying the contextual factors that influence teachers' use of pedagogical content knowledge gained through professional development is the first step for formulating strategies for supporting and sustaining teacher change.

4.2   Theoretical Framework

Teacher professional development is considered a valuable strategy for improving science education. Researchers report that effective professional development strengthens teachers' content and pedagogical knowledge and thereby promotes improved classroom teaching practices (National Staff Development Council 2001; Rotermund et al. 2017; Sparks 2002; Stigler and Hiebert 1999). Since elementary teachers have less extensive backgrounds in science than middle and high school teachers (Banilower et al. 2013; Olson and Labov 2009), finding ways to bolster teachers' knowledge and preparedness to teach science is particularly important in elementary schools. In addition to feeling less qualified to teach science than mathematics and language arts, elementary teachers report feeling less confident to teach science (Banilower et al. 2013; California Council on Science and Technology 2010; Dorph et al. 2011).

Desimone (2009) identified an operational theory of how professional development influences teachers, their instructional practice, and student learning. The core theory of action includes four main steps. First, teachers participate in effective professional development. Second, their participation increases their knowledge and skills or changes their attitudes and beliefs. Third, given their new knowledge and skills (and/or attitudes and beliefs), teachers adapt and improve their instructional practice through changes in content, pedagogy, or both. Fourth, these changes in instructional practices promote student learning. During each of these steps, context functions as a key mediating influence.

Our operational theory builds on Desimone's work by placing greater recognition of contextual influence on the entire process and includes both external and internal contextual factors. Contextual factors and organizational support have a significant influence on whether or not teachers implement new teaching strategies learned in professional development (Guskey and Sparks 2002; National Academy of Sciences 2015). Individual-level factors, such as teachers' knowledge and attitudes, as well as school-level factors, such as administrative support and resources, make a difference in teachers' decisions about changing their classroom instruction (Guskey 2002; Mumtaz 2000). To adopt teaching innovations, teachers must not only recognize the value of the instructional practice but also perceive that there is administrative support for the innovation (Sherry 2002). Administrative support not only affects the extent to which teachers initially change their instructional practices but also the extent to which they continue to use reform-based strategies (Banilower et al. 2007; Guskey and Sparks 2002; National Academy of Sciences 2015). Examining the influence over time of contextual factors on teachers who participated in similar professional development programs provides insight into the types of supports and barriers that may mean the difference between successful professional development outcomes (e.g., implementation of research-based instructional practices in science) and maintaining the status quo.

4.3   The Professional Development Programs

This study focused on K-6 teachers who voluntarily participated in one of four state-funded professional development programs. Each program ran approximately 3 years, in the period between 2007 and 2014. Each program included characteristics of well-designed, effective professional development (Darling-Hammond et al. 2009, 2017; Desimone 2009; Desimone and Garet 2015) including (1) sustained, intensive, and active adult-level instruction; (2) a focus on curriculum content rather than abstract principles; (3) a connection to the goals and priorities of participants' schools and districts; and (4) an emphasis on collegiality. All provided extensive opportunities for teachers to collaborate and included summer institutes and follow-up events during the school year. All were designed and implemented by highly experienced professional development experts, some of whom worked on all four programs. The authors of this chapter were not involved in designing or implementing these programs; rather, one author served as an external evaluator.

All of the programs aimed to increase teachers' pedagogical content knowledge in a variety of ways. First, each program was designed to prepare teachers to implement inquiry-based science strategies that were tied to the state science standards. Second, all of the programs emphasized the use of student notebooking to enhance student achievement in science. Specifically, during inquiry science lessons, students wrote about their predictions, the data they collected, and the conclusions they reached as a result of their inquiry. Thus, the professional development offered through these programs enhanced participants' pedagogical content knowledge by teaching them a new technique for helping students learn the scientific method. Third, during the professional development, teachers learned how to modify instruction to be grade-level appropriate. Consequently, even kindergarten and first-grade students could use notebooking and engage in inquiry science. Fourth, the professional development programs encouraged teachers to connect science to language arts and mathematics, which enhanced their pedagogical content knowledge in all three subject areas.

One difference between the programs relates to the grade-level teaching assignments of participants. One program was geared to teachers in the lower grades, another enrolled upper-grade teachers, and two had a wider range of grades represented. There were also curricular differences across programs. For example, one highlighted place-based science teaching, while another focused heavily on lesson study, and another emphasized environmental education.

4.4 Participating Teachers

One hundred thirty-five teachers from 31 schools in two dozen small districts in seven counties in Northern and Central California were in our sample. Seven of these districts are one-school districts. Student performance on standardized tests in these districts indicates a history of low academic achievement. A large percentage of students in these schools and districts are from economically disadvantaged families. Three of the four programs worked with teachers in schools with a high percentage of English Language Learners (ELL), many of whom are Hispanic.

4.5 Data Sources and Methods

Multiple sources of data and multiple methods were used to allow for triangulation of findings and provided a detailed picture of the participants' views about the impact of the professional development, the most critical supports they received, and the contextual factors influencing their science teaching. Specifically, data sources for this study included online teacher surveys, which were completed several times over the course of each program, pre-/post-science content tests specific to the content being addressed in the professional development, yearly interviews with a subsample of teachers and district/school administrators, yearly interviews with professional development staff, and yearly observations of select portions of the professional development over the course of each 3-year project.

One teacher survey (Horizon Research 2000) focused on teachers' opinions about science and science instruction, their preparedness, their instructional practices, and their perceptions of support for science instruction. The other survey—a self-efficacy assessment (Riggs and Enochs 1990)—focused on teachers' beliefs about their effectiveness in teaching science. Teacher interviews focused on teachers' perceptions of the impact of the professional development on their pedagogical content knowledge, and their choices of curriculum and instructional strategies.

Teachers also shared the extent to which they felt prepared to teach science using the techniques learned in the professional development, as well as the contextual factors influencing their teaching. Interviews of administrators focused on their awareness of and support for the professional development programs in which their teachers were participating; the impact they believed the professional development was having on teachers' instructional practices in science; and barriers and supports influencing science teaching at their school and in their district. Researchers interviewed project staff about their views regarding the success of the program in (a) enhancing teachers' pedagogical content knowledge, (b) creating teacher instructional change, as well as (c) contextual factors that might influence these changes. Finally, observations of the professional development provided informal opportunities to discuss what teachers were learning, as well as school and district contexts, with participants.

For analysis of survey data, researchers employed standard quantitative data analysis procedures, including preliminary descriptive checks for outliers, univariate, and cross-tabular analyses to check out-of-bounds and illogical values and analyses of missing data patterns. For the qualitative data, researchers used a combination of grounded theory (Strauss and Corbin 1998) and established methods for coding qualitative data (Miles and Huberman 1994). Researchers coded the data according to categories generated from the research questions, such as perceived impact of the professional development, instructional strategies, and contextual factors. In the contextual factors category, various sub-categories emerged from the data, such as time, administrative support, teacher support, and resources. Researchers organized the data by categories and sub-categories and then examined the data seeking corroborating and disconfirming evidence.

4.6   Individual-Level Teacher Outcomes

Survey, self-efficacy, interview, and observation data indicate that, across programs and across years, each professional development program had significant and positive impacts on individual-level teacher outcomes. All four programs were judged to be effective in terms of meeting the program goals of increasing teachers' pedagogical content knowledge, science content knowledge, and self-efficacy related to teaching science. On the Horizon teacher survey (2000), for example, pre-/post-results indicate that the majority of teachers in all four programs felt better prepared to use—and were more frequently using—research-based practices for teaching science, such as having students conduct experiments and engage in inquiry. Pre-/post-science content tests taken before and after each intensive professional development session consistently showed statistically significant improvement in teachers' scores. In each program, pre-/post-self-efficacy assessments (Riggs and Enochs 1990) indicated statistically significant, positive growth in teachers' confidence related to teaching science.

In interviews conducted each year of the project with a sample of teachers, participants voiced their enthusiasm about the professional development's impact and reported positive changes in their pedagogical content knowledge and their science content knowledge. Teachers felt that their ability to teach science curriculum—and their actual interest in teaching science—had changed significantly:

> This is…one of the most useful [professional development programs] I have ever taken in my over 17 years of teaching. Its broad scope and hands-on science teaching and the way the instructors taught us…[and] blended the content and teaching methodology into a useable, cohesive chunk of knowledge that is incredibly useable in my classroom.
>
> I have deepened my understanding of earth, life, and physical science directly because of this training. I am confident that I really know what I am talking about now and can go out and share information regarding science.

When asked how their instructional choices changed because of the professional development, teachers reported that their instructional focus shifted from emphasizing science facts to focusing on scientific processes, which indicated their enhanced pedagogical content knowledge. As one teacher noted, "I think that [the professional development] has reminded me in particular of the importance of these kids getting this science background and being critical thinkers and being a scientist and looking at things, and how important that curriculum is and that subject is." Students did more investigating and more frequently drove the direction of the lessons in class with their own interests and inquiry, according to teachers. This teacher's comment illustrates that teachers in these programs were taught to think not only of science content but also the ways in which science should be taught, which is clearly related to pedagogical content knowledge.

Another teacher described the instructional shift this way: "We've actually increased our hands-on learning because of [the professional development]. We spent more time planning. It's really brought science to the forefront, and made us try to make science more engaging." Teachers "[handed] the lesson and learning over to the students" by letting student interest lead the class. Lessons were much more "hands-on," and students spent more time working and talking in groups, collecting their own experimental data, and generating their own hypotheses and conclusions. As a result of these changes, both teachers and students were more interested in science. As one teacher reported,

> I actually look forward to when I teach science now. I actually had kids wanting to give up their recess so we could start science early. So, I think that says enough right there, because my kids will do anything to have recess, but were [asking], "Can we start now, and not do recess?" And I'm like, "Well, I won't stop you."

A significant element of the professional development was the support offered by project staff as well as other teachers in the program. For example, teachers reported that reassurance from project staff that it is "okay to try new things," "make mistakes," and "change or adapt lessons and try again" led to implementation of new teaching strategies in science. The professional development also incorporated ongoing opportunities for collaboration, which teachers highly valued. As one teacher from the program that focused on lesson study said:

> Sometimes, the best successes come by our errors. I know—I didn't personally teach the lesson … but it's one I observed [as part of lesson study]—where the first time the teacher

taught [the lesson], the data wasn't really what we wanted…. After she presented [the lesson], we collaborated, and then the next day when another teacher presented [the lesson], we changed it, and it was like, "That's how we do it." So, definitely with collaboration we were able to make the lessons better.

Each program sought to enhance pedagogical content knowledge by teaching content through modeling instructional strategies that teachers could take back to their own classrooms. Through these experiences, teachers saw first-hand that "having students create their own learning by finding the answers on their own" could have a positive impact. As one teacher described:

[Learners] definitely get so much more out of it when we do the inquiry first, and do the predictions about what's going to happen… And I appreciate [the professional development] because it's made me go through those steps, and not just read the book. [Learners] have to spend time, and learn to do their own inquiries, and to talk to each other. And whenever we do experiments [learned in the professional development]—and they take a long time, one and a half hours or more—they learn so much more. Usually it takes all morning to go through all the steps of the experiment, but it helps them remember. "Oh, remember that experiment we did?" vs. "Do you remember when we read that chapter and answered some questions?"

A teacher from a different program commented:

[Before the professional development], it was a lot more straight out of the book… Let's read this chapter, then answer some questions on a worksheet… Some activities, but not a whole lot… [Since the professional development], it's been a lot more student-led, just experimental kind of stuff… Getting to see the examples of notebooks [during professional development], it was just very eye-opening, how science can be approached differently.

4.7   Contextual Factors that Influenced Implementation

Contextual factors varied substantially across classrooms, schools, districts, and counties, which both fostered and hindered teachers' ability to use what they had learned in professional development. Although teachers learned science content, reported an increase in pedagogical content knowledge, valued what they learned, believed that they were capable of changing their instructional techniques, and felt that changing their science instruction would benefit students, factors such as access to resources, expectations about curriculum coverage, opportunities to collaborate, and administrative support had an impact on the extent to which teachers could successfully implement what they learned in the professional development. Data from these four programs indicate that what might matter most in terms of changing classroom instruction are contextual factors, rather than individual teacher change. Enhancing teachers' pedagogical content knowledge, for example, may be necessary for improving instruction, but it is not sufficient. In the following sections, we examine the most important contextual factors for these teachers: time for planning and collaboration, time for science instruction, administrator support, access to resources, and regional constraints.

4.7.1   Time for Planning and Collaboration

All four programs were designed to provide teachers with time to apply their new pedagogical content knowledge as they developed lessons using the new instructional strategies they were learning. The programs also gave teachers time for collaboration with peers. This collaboration not only introduced teachers to new ideas but helped them create a community of learners that participants and project staff hoped would last even after the professional development ended. In interviews, teachers noted being particularly impressed by the diversity and quality of their colleagues and finding the collaboration process built into each program generally "inspirational." Overall, teachers described the programs as providing a "good network" for support and ideas that were well suited to their current understanding of the science content and their grade level, as this quote illustrates:

> I think that the [professional development] participants do feel that there is a good network. I know people [from the project] call me or send me an email and say, "I heard that you did something with agriculture. Could you please send me a copy of that, or tell me where to get more information?" And I think that when teachers know there's a place that they can go to get information that's on their level so that it's not information that you have to wade through, and it's not information that someone is talking down to you, but it's just another teacher that's found a way to do something that they can get some help from, I think that's very helpful.

This type of support apparently was missing in many teachers' prior experiences. Teachers described how they initially operated essentially in isolation—running their classrooms as "an [independent] island,"—but were introduced to the benefits of collaboration by the professional development and found collaboration to be surprisingly helpful and time-saving. Teachers commented on the benefits of "getting together with the teachers and bouncing ideas off of each other," "creating lessons and talking to other teachers," and "brainstorming about the new things coming." They expressed concern that collaboration would be difficult after the professional development ended.

Although all teachers had time to collaborate within the structure of their professional development, schools and districts differed in the extent to which they provided additional time for teachers to meet. This challenge was especially true in elementary schools, which typically do not have built-in teacher planning periods. In some districts and schools, administrators managed to provide time for teachers to work together. One teacher, for example, commented:

> [Administrators] give us the time that we need to take off for different collaborations. Events during the year, or lesson studies, or different things. I am really lucky—at the site that I'm at, we have four different [teachers in the professional development program], so we have that collaboration constantly—not just when we go to meetings for it. The district really is just basically allowing us to have the time. And they also give us collaboration time—weekly grade level collaboration time—and they allow us to work on our [professional development] stuff.

At other sites, finding collaboration time was particularly challenging. In a number of schools, for example, severe budget shortfalls and increased requirements due to schools being designated as "Program Improvement" schools by the state led to restrictions on the amount of time that teachers could be released for collaboration related to science instruction. Program Improvement schools must improve student performance on standardized tests and, consequently, focus heavily on English Language Arts and mathematics. The project director felt this situation was not because of a lack of support from the principals, but was due more to circumstances beyond their control:

> [They are] experiencing severe budget shortages, and at the same time getting a lot more requirements of what they have to do. So, it's not like I think the principals wouldn't have liked to have [given teachers collaboration time]; it's that their hands are tied. It's kind of a situation where they're really caught.

Even when districts and schools gave teachers time for collaboration, logistical problems sometimes impeded teacher interaction. For example, some teachers across school sites collaborated on lesson study, a professional development opportunity that is particularly suited to enhance pedagogical content knowledge, but finding a mutually agreed upon time was often challenging:

> And she's the only one that teaches the after-school program at her school. So, scheduling the different things at the different school sites with schedules that were so diverse—that was hard. And with our school, finding subs with the different schedules at the different schools sites—we'd be ready to do it this week, but they would have open house that week. And it would be like, "OK, I can do it that week," but they would have a field trip that week. So just across the school sites was a little difficult.

4.7.2 Time for Science Instruction

Science has not been a regular part of daily classroom instruction in elementary schools for many years (Banilower et al. 2013; Dorph et al. 2011; McMurrer 2008; Olson and Labov 2009). Due to various accountability measures, administrators opt to devote more time and resources to mathematics and English language arts (Griffith and Scharmann 2008; Marx and Harris 2006; McMurrer 2008). Consistent with this research, teachers in these professional development programs reported that the greatest barrier to implementing what they had learned in professional development was the lack of time within the school day to teach science:

> The biggest barrier that I have with everything is always time. English Language strategies—you cannot shortcut them.... We don't have time to teach what we need to teach. I say we could easily extend the day by—I want to say an hour, but at least 30 minutes.
>
> As always remains the question—how do we get more time to incorporate all this great knowledge into the classroom?

Minimum days, library time, standardized testing, field trips, parent conferences, and other activities and events reduced available time for science, teachers explained. Even though the professional development improved teachers' pedagogical content knowledge, this improvement had little impact in those classrooms where teachers could not squeeze science into an already packed instructional day. Even when teachers were able to find some time to teach science, instructional strategies that are time-consuming—such as student notebooking or inquiry-based science—were often modified or, in some cases, abandoned in favor of lessons straight out of the science textbook. Again, school and district constraints minimized teachers' ability to apply the pedagogical content knowledge they had acquired in their professional development.

According to one project director, 50–75% of the teachers in the program "couldn't teach science as much as they wanted to." Moreover, teachers reported struggling to teach science as they had learned in the professional development program due to time constraints, both for planning and instruction:

> It's prep time, but it's [also] in-class time. It's when you get to the end of the period and you have to wrap it around to another day… When the end of the day comes and you still have more to do, but you can't. And there have been a couple of times where, sadly, I wasn't even able to get back to the lesson to wrap it up and conclude everything. I want to spend enough time to do it justice with my teaching, but, also, I'd rather conclude the lessons and have them completely done, so it's kind of a balancing act.
>
> The biggest challenge I had was finding more time—making more time to do it, and do it the correct way. And I know when I get strapped for time, I tend to pull back, and revert to old ways, and that's my biggest challenge: the day is not long enough.

Problems related to finding time for science were exacerbated in schools where teachers were required to use a highly prescribed program for ELL students, even though each program specifically taught teachers how to integrate science with English language arts and ELL instruction. As one project director explained:

> In some districts, they had a prescribed program that they had to follow, and they had to use the workbooks and all of that, and in some districts it was actually much stricter than others, and [administrators] would actually come in and check to make sure [teachers] were using the materials they were supposed to and on the page they were supposed to be on. So, it's really demoralizing for the teachers.

The lack of time for science instruction was also especially problematic in schools designated by the state as "underperforming." In many of these schools, pacing schedules for math and language arts and concerns about low student test scores in these content domains led to environments where science teaching was not valued or even expected. At the time that these professional development programs were implemented, science achievement was only a small part of California's accountability system. Moreover, since statewide standardized tests covering science did not begin until fifth grade, administrators pushed teachers in these schools to work on mathematics and language arts. As one teacher described:

> Pretty much the biggest barrier is time—time getting to science—because typically priorities are language arts and math, because those are areas where the [state] tests come in. Science is kind of third priority, because knowing that science is [tested in the] 5th grade… we do dedicate more time [for science] than we do for social studies.

When asked in an interview what she might change about this professional development program, the project director said:

> I guess what I would change is—and I don't know if that would be possible or not—but I would make sure that teachers were able to do all of the lessons that they wanted to do in

their classrooms. That was a real big issue for some districts because their [math and language arts] scores weren't high enough, so they had to follow a prescribed procedure from the district about what they could do and what they couldn't do. That made it really difficult for some of the teachers to do science, because they could have done their EL lessons with science, but they weren't allowed to.

In a few schools, teachers were allowed, and sometimes even encouraged by administrators, to integrate science with language arts to increase the time spent on science. Some teachers, for example, taught science during English Language Development (ELD), which they felt led to an improvement in students' vocabulary, as this quote illustrates:

The ELL learners did a lot of science during their ELD time. I felt like it was really important... It really most definitely helped. For example, if I took out all the magnets, you heard them speaking in English and saying things related to science like "attract," and they wouldn't have used that language without the science.

At another school, a participant also used ELD time to teach science. However, in her situation, students classified as lower level on the CELDT (California English Language Development Test) did not receive science instruction. She explained:

When our ELL kids have ELD, the advanced, early advanced, and English only students stay with me. So, [students classified as] CELDT 1, 2, and 3 don't get science.

One teacher (who was the only teacher at her school participating in the science professional development program) described her determination to make sure that all kindergartners at her school received science instruction. By using a rotating schedule and collaborating with other teachers at her school, all kindergarteners learned science "almost once a week." She stated,

We have a full-day kindergarten. At recess, I send my kids out and pull in another [teacher's] class, and [students] miss recess to have science.... There are six other classes that I work with. So, all the students at my grade level have science almost once a week. They are excited to come into class, and don't mind missing recess. We extend it to 35–40 minutes. Other teachers watch my kids.

Whether this rotating schedule would be feasible in kindergarten classes without a full-day schedule is an open question, but it is probably unlikely.

4.7.3   Administrator Support

Research on science teaching indicates that principal support predicts both instructional time in science and teachers' use of investigative strategies (Banilower et al. 2007). Knowing the importance of administrative support, the individuals who wrote the grant proposals discussed the professional development programs with school principals as well as district and county personnel to build support before the four programs were even funded. The hope was that these administrators would be more likely to support participating teachers at the school and district levels. This proactive approach appeared to work in some cases. For example, one teacher described how support from her district's Director of Curriculum led to "getting support from the district office, where they're saying, 'We'll give you collaboration time. Give us a list of materials for next month and we'll get those materials; tell us how you're going to be using them.' So, we have a lot more access than we ever did to materials and time to collaborate than we ever did."

However, even though administrators consistently *voiced* support for teachers' participation in these professional development programs, teachers did not always perceive their administrators as supportive. According to the project directors, district- and school-level support varied across sites. In some cases, the project directors and teachers described administrators who were "being supportive" as "not getting in the way," as this quote illustrates:

> So at [one district] we did have a lot of support at the site level by the principals. They didn't really get in our way, but they didn't really go out of their way to help or learn what we were doing. They didn't really attend any of our meetings. I think one principal [in this district] attended one meeting where we had a sharing-out of best practices and ideas, and we had called it Lesson Study Colloquium, and I think one of those principals attended one of those meetings … but that was it.

The situation was similar with administrators at the school level, according to one project director:

> At the site level, the principals there weren't very supportive either. It wasn't that they were unsupportive—it was that they were non-existent. They didn't really do anything to further the project, get to know what we were doing, observe the teachers in action, or anything like that.

In the views of the project directors, appropriate support from administrators would involve taking actions to "further the project," including getting to know what the teachers were trying to do, but these actions were not commonplace. At times, district or school support consisted of only one or two people who supported the program in various ways. Consequently, when these people left the district or school or were reassigned to other positions, the "support" for the program disappeared.

Another form of administrator support that was often lacking involved principals being willing to allow teachers to take instructional risks. As one participant described,

> You come in Year One, you have this lesson and you're really jazzed about [it] … and then you teach it and it bombs. If you're not part of a school that encourages risk-taking, then you may not try that again… You may not teach it when you think the administration's going to walk in the room. I mean, those are things that don't support success and growth.

In some schools, principal turnover made obtaining support for teacher change challenging. In one participating school, for example, there were three different principals in 4 years. As the project director explained,

> And the first principal who was at [one school in the project]…was probably the most supportive, but she was only there though the first intensive [professional development session in the first summer], and then when the second school year started up, she was

gone and there was another principal there. And then she lasted a year, and then there was another principal… I mean, it would have been really nice to have a principal there so they could see what was going on, and that they could really understand [the project]. I think that would have been really great, and I think there would have been a lot more buy-in that way.

4.7.4   Access to Resources

A consistent theme in interviews of teachers, project leaders, and administrators relates to the difficulties many schools face in obtaining resources for science teaching. Although a few participating teachers personally applied for and received science-related grant money during the projects to help fund science materials, most participating teachers described a significant lack of funding for science-related resources and activities. For example, at one school, the sixth grade science camp was eliminated due to a lack of funding, and there was a sense that the regional economy was too poor to ask parents to fund the camp themselves. Another teacher stated, "There are no funds for anything replaceable. We buy it ourselves. We had to pay for silkworm eggs." Some participating teachers managed to locate science resources that were not being used by others. One teacher, for example, described how she "scrounged around the school and [discovered] a lot of science materials no one was using." Another teacher said that the core science curriculum from her district gave her "a lot of what I need to do science experiments," such as manipulatives and consumables. Some teachers were able to borrow equipment from their county office of education, but others worked so far away from the office that they did not want to drive to pick up available materials. Even the teachers who had some resources consistently voiced concern in interviews about finding ongoing funding, since no amount of professional development could sustain hands-on instruction if teachers lacked the raw materials for these types of lessons.

4.7.5   Regional Constraints

Collegial support and collaboration are key factors for teachers to try out and implement new instructional strategies (Franke et al. 2001). In rural settings, teacher collaboration is even more important in reform efforts yet more difficult to arrange and maintain (Boyer 2006; Harmon et al. 2007). In the particularly small schools and districts in this study, teachers often had no grade-level colleagues or even other participating teachers for collegial interaction and support in planning and teaching science. The distance between the rural schools worked against regular collaboration with teachers in other schools. In some cases, the distance from the county office of education made it impractical for teachers to pick up and return supplies, such as science kits, that might be available for their use. In addition, falling enrollments in the districts sometimes added to the challenge of teaching science; classes were combined and included two or even three grade levels with one teacher. In these situations, teachers had to plan and implement curricula that would meet the standards of multiple grade levels. Understandably, this situation had an impact on both teacher planning and instruction in all content areas.

In contrast, an unusual circumstance in one very small one-school district proved to increase the time available for science instruction. The district decided to save money by switching to only one bus run for all K-8 students. This change in the bus schedule lengthened the instructional day for younger students who would normally be released at least an hour earlier than their older peers. In interviews, teachers in this particular district stated that having "extra" time in the day allowed them to teach science on a regular basis to all of their students.

4.8   Conclusion

In the 30 years since Shulman (1986) introduced the concept of pedagogical content knowledge, countless studies have been conducted about this concept with teachers at every grade level, in diverse subject areas—from science to history to engineering to statistics. A quick search of the literature reveals that the concept has been studied all over the world, from Cambodia to Turkey to South Africa to Iceland, with both pre-service and in-service teachers. Researchers have also expanded the concept; for example, Koehler and Mishra (2009) have introduced the concept of technological pedagogical content knowledge, which involves the interaction between pedagogical content knowledge and technology.

While the importance of understanding teachers' pedagogical content knowledge and its impact on instruction is clear, what is less apparent is the extent to which seemingly mundane factors, such as bus schedules or a lack of funds to buy consumable materials such as silkworm eggs, can impact the extent to which teachers apply what they have learned in professional development in their classrooms. Contextual factors, which vary greatly from school to school, district to district, and state to state, impact teachers' planning and their instructional priorities, which in turn influence the choices they make in their classrooms. As this chapter describes, teachers embraced what they learned in professional development, improved their pedagogical content knowledge, and wanted to implement inquiry-based science. However, despite their interest and motivation, contextual factors such as time for planning and collaboration, time for science instruction, administrator support, access to resources, and regional constraints had an important influence on their decisions related to science instruction. Given the cost of professional development, it is crucial to better understand the extent to which teachers can apply what they have learned and the factors that both support and constrain their instructional choices.

References

Banilower, E. R., Heck, D. J., & Weiss, I. R. (2007). Can professional development make the vision of the standards a reality? The impact of the National Science Foundation's local systemic change through teacher enhancement initiative. *Journal of Research in Science Teaching, 44*(3), 375–395.

Banilower, E., Smith, P. S., Weiss, I. R., Malzahn, K. A., Campbell, K. M., & Weiss, A. M. (2013). *Report of the 2012 national survey of science and mathematics education*. Chapel Hill, NC: Horizon Research, Inc.

Boyer, P. (2006). *Building community: Reforming math and science education in rural schools*. Fairbanks, AK: Alaska Native Knowledge Network.

California Council on Science and Technology. (2010). *The preparation of elementary school teachers to teach science in California*. Sacramento, CA: Author.

Darling-Hammond, L., Wei, R. C., Andree, A., Richardson, N., & Orphanos, S. (2009). State of the profession: Study measures status of professional development. *Journal of Staff Development, 30*(2), 42–50.

Darling-Hammond, L., Hyler, M. E., & Gardner, M. (2017). *Effective teacher professional development*. Palo Alto, CA: Learning Policy Institute.

Desimone, L. M. (2009). Improving impact studies of teachers' professional development: Toward better conceptualizations and measures. *Educational Researcher, 38*(3), 181–199.

Desimone, L. M., & Garet, M. S. (2015). Best practices in teachers' professional development in the United States. *Psychology, Society, and Education, 7*(3), 252–263.

Dorph, R., Shields, P., Tiffany-Morales, J., Hartry, A., & McCaffrey, T. (2011). *High hopes, few opportunities: The status of elementary science education in California*. Sacramento, CA: The Center for the Future of Teaching and Learning at WestEd.

Franke, M. L., Carpenter, T. P., Levi, L., & Fennema, E. (2001). Capturing teachers' generative change: A follow-up study of professional development in mathematics. *American Educational Research Journal, 38*(3), 653–689.

Griffith, G., & Scharmann, L. (2008). Initial impacts of No Child Left Behind on elementary science education. *Journal of Elementary Science Education, 20*(3), 35–48.

Guskey, T. R. (2002). Does it make a difference? Evaluating professional development. *Educational Leadership, 59*(6), 45–51.

Guskey, T. R., & Sparks, D. (2002). *Linking professional development to improvements in student learning*. Paper presented at the Annual Meeting of the American Educational Research Association, New Orleans, LA.

Harmon, H., Gordanier, J., Henry, L., & George, A. (2007). Changing teaching practices in rural schools. *The Rural Educator, 28*(2), 8–12.

Horizon Research, Inc. (2000). *Local systemic change through teacher enhancement science teacher questionnaire*. Chapel Hill, NC: Horizon Research, Inc. http://www.horizon-research.com/instruments/lsc/tq_k8sci.php.

Koehler, M.J., & Mishra, P. (2009). What Is technological pedagogical content knowledge? *Contemporary Issues in Technology and Teacher Education (CITE), 9*(1), 60–70.

Marx, R. W., & Harris, C. J. (2006). No Child Left Behind and science education: Opportunities, challenges, and risks. *The Elementary School Journal, 106*, 455–466.

McMurrer, J. (2008). *Instructional time in elementary schools: A closer look at the changes for specific subjects*. Washington, DC: Center on Education Policy.

Miles, M. B., & Huberman, A. M. (1994). *Qualitative data analysis: An expanded sourcebook* (2nd ed.). Thousand Oaks, CA: Sage.

Mumtaz, S. (2000). Factors affecting teachers' use of information and communications technology: A review of the literature. *Journal of Information Technology for Teacher Education, 9*(3), 319–342.

National Academy of Sciences. (2015). *Science teachers' learning: Enhancing opportunities, creating supportive contexts*. Washington, DC: The National Academies Press.

National Staff Development Council. (2001). *Standards for staff development (Revised)*. Oxford, OH: National Staff Development Council (NSDC).

Olson, S., & Labov, J. (2009). *Nurturing and sustaining effective programs in science education for grades K-8*. Washington, DC: The National Academies Press.

Riggs, I. M., & Enochs, L. G. (1990). Toward the development of an efficacy belief instrument for elementary teachers. *Science Education, 74*, 625–637.

Rotermund, S., DeRoche, J., & Ottem, R. (2017). *Teacher professional development by selected teacher and school characteristics: 2011-12*. Washington, DC: U.S. Department of Education.

Sherry, L. (2002). Sustainability of innovations. *Journal of Interactive Learning Research, 13*(3), 211–238.

Shulman, L. (1986). Those who understand: Knowledge growth in teaching. *Educational Researcher, 15*, 4–14.

Sparks, D. (2002). *Designing powerful professional development for teachers and principals*. Oxford, OH: National Staff Development Council.

Stigler, J., & Hiebert, J. (1999). *The teaching gap*. New York: Free Press.

Strauss, A., & Corbin, J. (1998). *Basics of qualitative research: Techniques and procedures for developing grounded theory* (2nd ed.). Thousand Oaks, CA: Sage.

Chapter 5
Teacher Knowledge and Visual Access to Mathematics

Jill Neumayer DePiper and Mark Driscoll

Abstract We propose that there exists mathematical knowledge for teaching (MKT) specific to visual representations (VRs), abbreviated MKT-VR. We define a VR as a graphic creation, such as a diagram or drawing, which illustrates quantities and shows quantitative relationships or which illustrates geometric properties of figures and shows geometric relationships. A teacher with strong MKT-VR will, for example, be able to use and understand VRs in his/her own problem solving and will have mathematical knowledge specific to teaching students to use, analyze, and solve problems with VRs. The Visual Access to Mathematics (VAM) project seeks to help teachers understand the value of VRs, specifically when teaching and learning ratio and proportional reasoning content. This chapter lays out a theoretical framework that we anticipate using to guide and benchmark future research.

Keywords PCK · Mathematical knowledge for teaching · Visual representations · Student thinking · Teacher knowledge · Proportional reasoning

5.1 Introduction

The chapter describes a construct of mathematical knowledge for teaching (MKT) specific to visual representations (VRs), abbreviated MKT-VR, focused on the context of ratio and proportion in particular. We detail why MKT-VR related to ratio and proportional reasoning is important for middle grades mathematics teachers, and how it relates to what is important for students. Following our discussion of the importance of MKT-VR, we describe our study on supporting teachers' MKT-VR in teacher professional development, and how we are measuring teacher MKT-VR related to ratio and proportional reasoning. Beyond the topic of ratio and proportional reasoning, we argue for an interest in studying MKT-VR across school mathematics. This would be consistent with growing evidence of the instructional

J. N. DePiper (✉) · M. Driscoll
Education Development Center, Inc., Waltham, MA, USA

© Springer International Publishing AG, part of Springer Nature 2018
S. M. Uzzo et al. (eds.), *Pedagogical Content Knowledge in STEM*,
Advances in STEM Education, https://doi.org/10.1007/978-3-319-97475-0_5

efficacy of visual representations for all mathematical content from kindergarten into high school.

5.2 Why Proportional Reasoning?

Middle grade mathematics sits at the crossroads in a student's school mathematics journey: it marks the end of focused attention on number and operations, and it comes before other themes, such as geometry and measurement, which are explored in depth in high school. While middle grades mathematics classes are not always organized around a theme, proportionality is an overarching concept in these grades, and "one that unites, relates, and clarifies many important middle grades topics" (Lanius and Williams 2003, p. 392). Proportionality has been called the "cornerstone of higher mathematics and the capstone of elementary concepts" (Lesh et al. 1988, p. 98). As the capstone to elementary topics, proportionality as a concept builds from understandings about number and operations and invites connections to real-world situations and to variation. In mathematics in high school and beyond, during more in-depth studies of algebra, probability, geometry, and measurement, understanding of proportionality is a prerequisite, as relationships between quantities are key to functions and variation. Proportionality also has many important connections outside of mathematics. Proportional literacy serves all citizens well, whether it is by using unit rates to compare grocery prices, understanding the relevance of growth rates in measuring economic health, or being aware of how population proportions influence political decisions.

Proportional reasoning, including the study of proportionality, refers to a mathematical way of thinking, specifically: "in which students are solving problems about proportional situations: Proportional *reasoning* refers to detecting, expressing, analyzing, explaining, and providing evidence in support of assertions about proportional relationships" (Lamon 2007, p. 647). Reasoning proportionally does not mean just using a cross-product approach for solving problems but includes understanding *when* problems are proportional or not and *what* the situations mean. Proportionality may be illustrated: in descriptions of how quantities vary, such as "Mark earns \$85 every two weeks..."; algebraically, as in linear functions, $y = mx$; or geometrically, as a line that passes through the origin (Lanius and Williams 2003). Also foundational to proportional reasoning, students need to understand ratios both as *composed units*—e.g., that in the ratio 1:4, for every one unit of one quantity, there are four units of another quantity—or as *multiplicative comparisons*, where there are four times as many of the second quantity as the first (Lobato and Ellis 2010). Understanding proportionality, its applications, and how to reason proportionally are foundational to middle grades' students' further studies, and attention to these areas can engage students in key mathematical content and practices.

5.3 Tackling the Complexity of Proportional Reasoning with Visual Representations

Using visual representations (VRs) can support learners' understanding of these key mathematical concepts and practices. VRs are a graphic creation, such as a diagram or geometric drawing, which illustrate quantities and/or geometric properties and show relationships among quantities and/or geometric figures[1]. Examples of VRs in rational number and proportional reasoning contexts are number lines and rectangular tape diagrams (sometimes called "strip diagrams"). Research specifically recommends VRs to reinforce students' conceptual understanding of rational numbers (Gersten et al. 2009; Siegler et al. 2010). A VR can call students' attention to the quantities presented in a problem and the relationships *between* quantities. VRs can also scaffold students' understanding of the symbolic approach to the problem (e.g., Siegler et al. 2011), through, for example, presenting equipartitioning and highlighting relations among fractions, decimals, and percentages (Gersten et al. 2009).

As a tool, VRs can support problem solving, communication, and engagement. Using a VR can support students in making sense of the problem, and then subsequently make modifications in light of sense making, and select a solution strategy (Ng and Lee 2009). Visual representations can provide a bridge from text to arithmetical or algebraic representations, which is a valuable support for all students and invaluable for students who are English Learners. VRs help students by linking the relationships between quantities in the problem with the mathematical operations needed to solve the problem. Opportunities to use mathematical visual representations provide students access to mathematics, support their engagement in problem solving, facilitate communication of their mathematical thinking, and develop the mathematical practices outlined in the CCSSM.

More generally, VRs are an element of multimodal communication. Multimodal mathematical communication refers to the various ways in which students convey their mathematical thinking, including language, gestures, drawings, or the use of tools (e.g., physical models, manipulatives, and technology). Students may use a combination of modes at once or different ones in isolation. To enhance mathematical learning opportunities for all students, particularly ELs and those struggling with language, research stresses the importance of creating classroom environments that encourage multimodal communication (Chval and Khisty 2001; Khisty and Chval 2002; Moschkovich 2002). Such environments can foster the development of the Standards for Mathematical Practice (SMP) that are articulated in the Common Core State Standards for Mathematics (CCSSM, CCSSM SMP; NGA & CCSSO 2010) by providing students access to the mathematics, helping them construct viable arguments, and providing opportunities to attend to precision by developing accurate mathematical language.

---

[1] This definition of a mathematical visual representation is particularly germane to number and algebra contexts.

A VR can also be a powerful tool toward eliciting the use of thinking aligned with the Standards for Mathematical Practice (SMPs). As much of middle-grade mathematics relates to quantities, spatial properties, and related problems solving, VRs can support learners in identifying and making sense of these quantities, properties, and relationships in ways that align with SMP 2 (*reason abstractly and quantitatively*) and SMP 7 (*look for and make use of structure*). Quantitative reasoning involves reasoning about the relationships among quantities and does not necessarily need to involve algebraic expressions or assigning variables (Smith and Thompson 2007). SMP 2 emphasizes rich conceptual understandings and not reliance of algorithmic thinking: "Quantitative reasoning entails habits of creating a coherent representation of the problem at hand; considering the units involved; attending to the meaning of quantities, not just how to compute them" (CCSSI 2010, p. 6). SMP 2 specifically emphasizes the ability to decontextualize and contextualize when using mathematics to solve problems (NGA and CCSSO 2010), and a VR can present mathematics in ways that a symbolic representation does not. For example, suppose a word problem says that *Maria has $10 more than Albert, and together they have $40.* We can write a symbolic representation of this situation, say, $M + A = (A + 10) + A = 40$, with M and A representing, respectively, Maria's amount and Albert's amount. In doing so, we have *decontextualized* the situation, that is, abstracted the quantitative information. VRs can represent this same information, support the decontextualizing and recontexualizing, and provide an artifact for that conversation. Using structure (SMP 7) is emphasized in proportional reasoning contexts, such as when learners use a double number line to note how, as two proportional quantities covary, the ratio between them remains invariant.

5.4 Student and Teacher Understanding of the Importance of Visual Representations

While VRs and opportunities to use them in mathematical thinking and learning can support mathematical reasoning, learning how to use a visual representation as a tool is a skill, in and of itself. Unlike geometry tasks, tasks where quantitative reasoning is prominent, such as algebraic word problems, usually do *not* provide visual representations. In such cases, knowing how to draw one's own visual representations is a very valuable skill. The 2012 IES Practice Guide, *Improving Mathematical Problem Solving in Grades 4–8*, based on an examination of hundreds of relevant, rigorous studies, recommends teaching students how to use VRs to enhance their mathematical problem solving: "Students who learn to visually represent the mathematical information in problems prior to writing an equation are more effective at problem solving" (Woodward et al. 2012, 23). It is important to study and become proficient with a variety of visual representations and to understand how to select the representations most appropriate for solving a task (Woodward et al. 2012). This "includes knowing what particular representations are able to illustrate or explain, and to be able to use representations as justifications for other claims" (Zbiek et al. 2007, p. 1192). Furthermore, the importance of rich understanding of VRs is not limited to middle school students, as research has found that competent mathematical thinkers, in university-level mathematics, use VRs flexibly in problem solving (Stylianou 2002; Stylianou and Silver 2004).

The ability to interpret and construct various mathematical representations, and to change representations appropriately, is considered *representational fluency* or *diagram sense*. A learner needs to develop diagram sense—knowing when various VRs are most useful—for example, knowing that tree diagrams can help organize probabilistic thinking, that number lines are often handy for rational number tasks, and that tape diagrams can propel thinking about algebraic word problems. We know that learners need to develop number sense in order to judge the reasonableness in their own and others' calculations. We also know that number sense can be learned through ample opportunities to reason about numbers and operations. Similarly, we believe that *visual representation sense* can be learned through ample opportunities to represent problems and to reason with the diagrams.

A critical piece in supporting students' representation fluency and fostering a culture of multimodal communication and engagement in the classroom in these ways is for teachers to value and encourage the use of mathematical VRs, especially as tools for reasoning and communicating mathematical thinking. For example, students need a diet rich with representations and an opportunity to study a variety of VRs to be able to understand how to select the representations most appropriate for solving a task (Woodward et al. 2012). In our experiences across multiple studies on teacher instruction, however, we have observed that students' opportunities for reasoning with and about quantities and relationships too often take a back seat to opportunities for students to practice computational procedures. The complex nature of proportional relationships is not consistently tackled in instructional situations. Without attention to conceptual understanding of relationships, learners may struggle to identify rates and ratios, reason about variation, or follow changes in units or analysis. Students may rely on algorithmic thinking, such as cross products, and not engage in reasoning proportionally, which does not strengthen their understandings of these key ideas.

Research has found that teachers may be less prepared in some mathematics content, as compared to other content areas, and specifically less prepared in the key areas of fraction and proportional reasoning and representing relationships (Siegler 2011). In reasoning proportionally, teachers tend to rely on algorithmic thinking, such as the cross-multiplication algorithm in proportional situations (Orrill and Brown 2012; Riley 2010; Singh 2000), and may be using cross multiplication as a procedure without conceptual understanding. Teachers often focus student attention on operational understandings and algorithmic thinking, instead of developing their conceptual understandings of proportional reasoning (Lamon 2007). Lack of understanding of content may limit teachers' ability to teach critical mathematics content: "Researchers typically do not associate reasoning with rule-driven or mechanized procedures, but rather with mental, free-flowing processes that require conscious analysis of the relationships among quantities" (Lamon 2007, p. 647). Teachers need a rich understanding of mathematics content, many times noted as *specialized* mathematical knowledge for teaching (Ball et al. 2008; Ma 1999), and also understandings of visual representations in proportional reasoning, similar to how visual representations support student learning in these areas.

Evidence from our previous research from the IES-funded project, Mathematics Coaching for Supporting English Learners (MCSEL)[2] and from others (Stylianou 2011) suggests that US teachers in elementary and middle grades generally are neither experienced nor skilled in understanding and using VRs in mathematics. From analysis of teacher instruction and self-reflection, we identified a critical need in continuing to build teachers' knowledge of how to use diagrams in their own mathematical problem solving. MCSEL developed and studied professional development (PD) for middle grades mathematics teachers of ELs that emphasized the use of visual representations integrated with language support strategies. As we introduced instructional activities to support students in using a visual representation when they approach tasks, we found that teachers also needed support in learning how to use visual representations in their own mathematical problem solving. Related research has also found that while teachers had important knowledge about proportions, their understanding of representations was not coordinated with their understanding of proportions (Orrill and Brown 2012). Participants in teacher professional development were found to rely on addition and subtraction strategies, rather than multiplicative reasoning, and initially struggled with a double number line representation (Orrill and Brown 2012). This research supports these findings, and along with other literature, this suggests that teacher professional development needs to focus on building connections between representations and ratio/proportional reasoning concepts and content.

Finally, we have also been struck by a common perception about VRs exhibited both by students and teachers—namely, that the primary purpose of a VR is to present the *product* of one's thinking about a mathematical task. No doubt this is a helpful role for VR, but it is far from the only, or even most valuable, purpose of creating a VR (e.g., Stylianou 2011). Rather, VRs can be really effective *reasoning tools* for learners trying to solve challenging mathematical tasks. This is a quality that makes them especially valuable for ELs. VRs are also *communication tools*, as students can use them to share their thinking and reasoning, and teachers can use students' VRs to prompt them to discuss relationships between quantities. VRs, far from being just a product of student thinking, can be part of the process of mathematical thinking, reasoning, and communicating.

[2] The Mathematics Coaching for Supporting English Learners research was supported by the Institute of Education Sciences, US Department of Education, through Grant R305A110076 to the Education Development Center, Inc. The opinions expressed are those of the authors and do not represent views of the Institute or the US Department of Education.

5.5   Teacher Professional Development Focused on VRs and Proportional Reasoning Content

The Visual Access to Mathematics (VAM) project seeks to help teachers understand the value of VRs, specifically when teaching and learning ratio and proportional reasoning content. The NSF-funded VAM project[3] is a multi-year design and development project that includes the development, facilitation, and related research on teacher professional development. It seeks to advance knowledge in the field about supporting mathematics teachers of students who are English Learners (ELs) by developing and studying a 60-hour blended-learning professional development (PD) program for middle-grade mathematics teachers who teach ELs. Concentrating on ratio and proportion and related rational number concepts, and relying heavily on technology-supported artifacts of student thinking, the VAM professional development (VAM PD) helps mathematics teachers of ELs become better at making, using, and analyzing VRs for mathematical problem solving, with the goal of improving teacher knowledge and practice. Embedding opportunities for students and teachers to use VRs in classroom environments in combination with teacher attention to and use of students' thinking to create equity in mathematics instruction maximize the value of VRs. The VAM PD blends online and face-to-face components; it includes a summer institute, eight online sessions, and two face-to-face workshops. Both online and face-to-face sessions include activities that focus on developing teacher knowledge in four areas: teacher knowledge of visual representations for problem solving, mathematical knowledge for teaching ratio and proportional reasoning, analysis of student mathematical thinking, and instructional planning with visual representations and language access and production strategies.

As we develop the VAM PD, we are also studying if and how it supports teacher knowledge, analysis of student work, and instructional planning. We hypothesize that developing teachers' abilities to use VRs for problem solving can improve teachers' mathematics instruction and provide ELs greater access to learn productive mathematical reasoning. Two overarching questions guiding this work are: (1) What supports will allow mathematics teachers to develop the pedagogical content knowledge they need to support ELs in mathematical problem solving? and (2) What is the effect of VAM PD on teachers' pedagogical content knowledge about using VRs to support mathematical problem solving? Key elements of the teacher intervention, teacher outcomes, and classroom/student outcomes are presented in the Theory of Change (Fig. 5.1). The focus in this PD and the related research is on the relationship between the VAM intervention and the teacher outcomes, as noted by the bold arrow in Fig. 5.1.

[3] The Visual Access to Mathematics project is supported by the National Science Foundation under Grant No. DRL 1503057. Any opinions, findings, and conclusions or recommendations expressed are those of the author and do not necessarily reflect the views of the National Science Foundation.

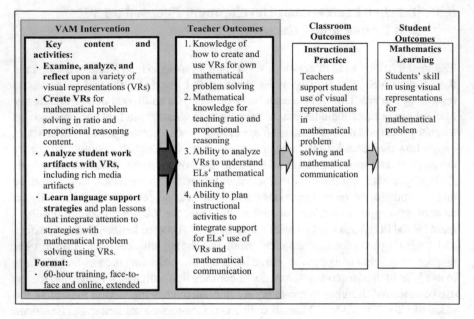

Fig. 5.1 VAM theory of change

5.6 Defining Teacher Knowledge of VRs

In the VAM project, there are multiple ways in which we are seeking to understand the relations between participant experiences and their shifts in knowledge and practice. Specific to our goals of measuring teacher knowledge, we argue that there is a body of knowledge and set of skills associated with fluent use of VRs in mathematics learning and teaching, and we posit that VAM PD will promote such knowledge and skills (Teacher Outcome 1). Our two different target sub-outcomes related to Teacher Outcome 1 (*Teacher Knowledge of Visual Representations*) further specify our interests in teacher learning: (a) improved the ability to use VRs to represent and solve ratio and proportional reasoning problems and (b) improved the ability to evaluate the strengths and limitations of different solutions involving VRs. It is important to revisit the rich educational research in the area of teacher knowledge as we define what teacher knowledge is as related to visual representations.

Thirty years ago, it was groundbreaking when Shulman (1986) placed a focus on the critical role of content knowledge in pedagogy, defining pedagogical content knowledge and identifying major categories of teacher knowledge that were content specific. As many others have expanded on this work, in mathematics education, Ball and colleagues (2005, 2008) have sought to define the knowledge needed by teachers, in order to better understand, measure, and support teacher knowledge. It is their definition that aligns with our current attention to teacher knowledge of VRs. Their working definition of mathematical knowledge for teaching is "the mathematical knowledge that teachers need to carry out their work as teachers of mathematics" (Ball et al. 2008, p. 4).

This attention to the *work* that teachers do as teachers is key in our understanding of teacher knowledge of VRs. As we emphasized in our earlier description of the importance of VRs, VRs support learners' conceptual understandings of content, and understanding in mathematics contexts is content knowledge. As teachers need to know and understand VRs enough to be able to teach them to others and VRs support and are a tool for mathematical thinking and communication, then teachers' VR knowledge is as much a tool for their own problem solving as it is a tool that teachers should teach with and teach their students to use.

5.6.1 Components of MKT-VR

Our VAM team theorizes that teachers need mathematical knowledge for teaching related to visual representations (MKT-VR), which includes both content knowledge and pedagogical content knowledge, situated in the *work* of teaching of middle grades mathematics. In extending the Ball et al. (2008) framing of mathematical knowledge for teaching (MKT) to teaching and learning with VRs, we suggest that a teacher with MKT-VR will be able to use and understand VRs in his/her own problem solving and will have mathematical knowledge specific to teaching students to use, analyze, and solve problems with VRs. A teacher with strong MKT-VR will integrate VRs into their teaching, will be able to generate inferences about students' thinking from students' VRs, and will consistently consider how to promote student use of VRs. Teachers need MKT-VR to conduct tasks such as identifying correct solutions and solving mathematical problems for themselves. Teachers also need MKT-VR, which includes being able to provide students with explanations for why particular solution strategies work, to diagnose student errors with different strategies and to understand nonstandard yet effective problem solving.

We believe that MKT-VR includes both content knowledge and pedagogical content knowledge related to VRs, and it can be defined broadly as a *body of teacher knowledge and set of skills associated with fluent use of VRs in mathematics learning and teaching*. While other models of MKT (Ball et al. 2008) further subdivide content knowledge and pedagogical knowledge into several subdivisions, at this time, we do not try to mirror these categories in our definition of MKT-VR. Instead, we seek to define the construct, and determine which understandings are key as related to VRs. CK-VR and PCK-VR are distinct but related elements of MKT-VR and related elements of middle-grade mathematics teacher knowledge. We theorize three important categories of MKT-VR, two within the area of content knowledge (CK) and one focused on pedagogical content knowledge (PCK), and we will refine these as we continue to learn from work with participants in the VAM PD.

5.6.2   Content Knowledge Related to Visual Representations (CK-VR)

CK-VR is mathematics knowledge related to classroom instruction, students or teaching, specific to visual representations. While content knowledge for teaching is broader than CK-VR, we identify specific types of content knowledge specific to visual representations. As teachers in our work have engaged with VRs for their own mathematical learning and in classroom instruction, the importance of *alignment* of VRs to their purposes and the importance of *strategic use* of VRs in their work have emerged as two key areas of content knowledge. *Alignment* includes making connections between the mathematics task being considered and the solver's purpose (process or product), where there is attention to alignment with the nature of the task and alignment with purpose for using the VR. *Strategic use* is knowledge about *how* to use a VR in situation.

Alignment   VRs can play multiple roles as a tool in problem solving and for communicating mathematical concepts and solutions with others (Stylianou 2011). VRs may support individual cognition by organizing information, recording information and reducing cognitive load, allowing manipulation of information and therefore facilitating exploration, and supporting the problem solver in monitoring progress and approaches (Stylianou 2011). Visual representations can also support communicating mathematically and be a tool in social practice for presenting obvious and not-so-obvious perspectives and information and for allowing the sharing of strategies and negotiation of new ideas (Stylianou 2011). Because of these multiple roles that VRs can play in mathematical problem solving, key knowledge for teachers is an understanding of which VR to use when and how to align a VR to a goal. The goal may be focused on specific mathematics content, such as related to supporting or conveying understanding of unit rates, or it may be about communicating mathematically.

Strategic Use of VRs   While a VR can serve multiple roles in mathematical problem solving, all VRs may not be equally effective. In our previous IES-funded MCSEL research, we identified four features of strategic use of VRs for mathematical problem solving[4]:

- Clear representation of the given *quantities* in a problem.
- Clear representation of the *relationships between the given quantities*, including surfacing implicit relationships or attending to proportionality. In ratio and proportional reasoning contexts, given quantities and relationships can be represented on a tape diagram and a double number line, and new quantities and relationships can be added to, or emerge, on a tape diagram and on a double number line to help solve the task.

---

[4] Here we describe VRs in quantitative settings. For geometric settings, comparable wordings refer to geometric properties and relationships, instead of quantities and quantitative relationships.

- *Strong potential to reveal relationships visually between the given quantities and the goal quantity (connecting a VR to an algebraic or symbolic relationship).* Effective VRs highlight connections between VRs and related symbols/calculations/algorithms.
- *Identification within the VR of features related to important mathematical concepts leading to the* goal. The important concepts could include unit rate or multiplicative comparisons.
- *Clear labels and markings* (such as shading, labels, dotted lines, etc.) that support managing data, recording the flow of reasoning, and representing or communicating the quantities and relationships in the problem.

We argue that the alignment of a VR to the purposes listed above (i.e., understanding information, recording information, supporting exploration and problem solving, presenting perspectives, and sharing strategies) is enhanced through using a VR strategically with attention to the features listed above. For example, it is more strategic to represent quantities clearly and highlight implicit relationships when VRs are used to organize and record information. When VRs are used to share strategies and ideas with others, labels and shading as well as attention to scale in the drawing may be important features of a VR to support its use as both a communication tool and a problem solving tool. A VR that identifies a unit rate or the relationships between quantities may be a strategic use of a VR to support connections to algorithms or calculations and may align with mathematical goals.

These two aspects of MKT-VR are primarily CK-VR, though their implications are not strictly restricted to content. For example, a teacher response (VAM teacher participant, Session 4, November, 2016) presents how content knowledge and pedagogical knowledge intertwine in the areas of *alignment* and *strategic use*:

> I love the idea that one can use two tape diagrams stacked on each other to make concrete visual comparisons of quantities, such as what we observe for the mixing paint tasks. ...For the mixture problems, the other key idea to take away is that both tape diagrams must be the same size! (Like a common denominator...

In the response, the teacher appreciates that one can use the area in tape diagrams to compare paint mixture proportions, and notes alignment among task, content, and VR. Specifically, if the task is to decide if 4 parts blue paint mixed with 3 parts yellow paint is darker than a mixture of 5 parts blue to 4 parts yellow, one can divide one tape in a 4:3 ratio and another tape in a 5:4 ratio and see which tape has proportionally more blue. The teacher adds that, to make this area comparison work, the two tapes need to be congruent, so there is a common unit guiding the measures. The response attends to why and how to use a VR and one with a common unit, therefore highlighting strategic use. We hypothesize that VAM PD will promote both types of CK among all teachers: knowledge of the alignment of VRs and how to use a VR strategically in mathematical problem solving.

Pedagogical Content Knowledge Related to Visual Representations (PCK–VR) We argue that a teacher with strong MKT-VR will have not only strong CK-VR but also strong pedagogical content knowledge related to VRs (PCK-VR). PCK includes knowledge of common student errors and misconceptions (or knowledge of students and content); mathematical models, representations, and contexts commonly used by or for students; and the ability to address and understand students' interpretation of mathematics (Ball et al. 2008; Campbell et al. 2014; Hill et al. 2008). We propose that PCK-VR includes these same elements, with an emphasis on VRs. For example, a teacher with high PCK-VR will have strong knowledge of how students think about or learn to use VRs and knowledge of how to teach problem solving with VRs.[5]

5.6.3 Measuring and Assessing MKT-VR

We define MKT-VR to include the sub-constructs CK-VR and PCK-VR, and we have two assessments to measure it. While the VAM project, in which the work to define this construct is currently embedded, does not include assessment development as part of its scope, we are nevertheless developing two exploratory assessments, described below. Assembling and developing our own assessments was necessary as we did not find any available assessments that focused on middle-grades teacher knowledge of visual representations that are appropriate given the scale of this study, where more than 100 teachers will participate in field testing of the PD. For example, previous research on mathematicians' and college students' understandings of VRs (Stylianou and Silver 2004) used interviews and comparisons of VR use between experts and did not focus on the content knowledge needed for the *work* of teaching. While research that involved interviews with teachers solving mathematics tasks with visual representations has provided information on how teachers use and understand VRs (Stylianou 2002), interviews are not appropriate for the scale of our study, and the interviews were not as narrowly focused on teacher knowledge and skills as we are in this work. It is also important to note that research on the importance of VRs for students' problem solving focuses more on the student's correct responses to tasks than it does on mathematical reasoning or communication (Boonen et al. 2014). Boonen et al., for example, make the distinction between a pictorial representation, an inaccurate visual-schematic representation, and an accurate visual-schematic representation. A distinction between a pictorial diagram and the visual-schematic may be too simplified for our purposes, as our participants are drawing diagrams that represent the mathematics (not pictures of the problem context), and we seek to understand the nuances in their ability to do so. In addition, as we are interested in supporting teachers and students in

---

[5] Ball et al. (2008) write that pedagogical content knowledge may be subdivided into several separate domains, including knowledge of content and students, knowledge of content and teaching, and knowledge of curriculum. In our definition of PCK-VR, we have not tried to define similar sub-constructs. Instead, we have tried to identify and include within a general PCK-VR construct the most important types of pedagogical content knowledge that teachers may need to teach mathematics with VRs effectively.

engaging in the mathematics and gaining access to the task, we recognize that pictorial representations have value for supporting access and language production, and could serve as a tool for student or teacher learning, and thus would treat teachers' knowledge about pictorial representations differently. Given this mismatch in purpose and fit between existing measures related to VRs and our endeavor to understand MKT-VR, we have decided to create two exploratory assessments.

The MKT-VR multiple-choice assessment is based on existing instruments and their items and seeks to measure target outcomes of VAM PD. To assemble the assessment of teachers' MKT-VR, the team has followed general scale development guidelines suggested by DeVellis (2012). Due to project constraints, the team restricted the scale's format to multiple-choice items and used items from established assessments of MKT rather than write original items. After the team developed a list of key skills that the assessment would target, the team reviewed existing test banks and assessments and selected an item pool of 30 multiple-choice items that could potentially measure these skills. Items came from the following sources: the *Learning Mathematics for Teaching* assessment (2009); research by DePiper et al. (2014); the National Assessment of Educational Progress (NAEP, grades 4 and 8); released items from the Educational Testing Service's Praxis tests (focusing on elementary and middle grades mathematics); and released items from the Massachusetts Comprehensive Assessment System (MCAS). Because items were drawn from different sources, the number of distracters associated with each item ranges from 3 to 5. The team tried to generate redundancy of items to measure individual skills (DeVellis 2012).

As there are limits in understanding teacher thinking when using a multiple-choice format instead of an open-response assessment, we have also developed an assessment that is an open-response analysis of VRs that are provided to the respondent. We have other measurement tools within the VAM research plan that will measure other types of teacher knowledge of student thinking and teacher practice in classrooms, related to Teacher Outcomes 2–4 in the Theory of Change (see Fig. 5.1), which will provide different information about teacher PCK-VR and will complement the findings from these multiple-choice and open-response assessments of MKT-VR.

MKT-VR Multiple-Choice Assessment   We assembled an assessment using multiple-choice items from other established instruments to measure both of these subconstructs, CK-VR and PCK-VR. The goal of analysis of the assessment responses will be to draw inferences about teachers' abilities to make and use effective VRs in their own rational number problem solving (CK-VR) and to use VRs for teaching mathematical problem solving as well as for understanding students' mathematical thinking (PCK-VR). We seek to draw inferences about CK-VR and PCK-VR as separate sub-constructs and about MKT-VR as a whole.[6]

---

[6] Our overall project hypotheses are the following: Compared to control group teachers, teachers who participate fully in VAM PD will demonstrate higher (1) MKT-VR, (2) MKT related to ratio and proportion, (3) ability to analyze student thinking, and (4) ability to plan lessons and activities that integrate VRs and language strategies in mathematical problem solving. The MKT-VR assess-

Fig. 5.2 Sample MKT-VR item. (MTEL 2011)

Given that the points on each of the number lines shown below are equally spaced, on which of the following number lines does point D correspond with the fraction $\frac{1}{4}$?

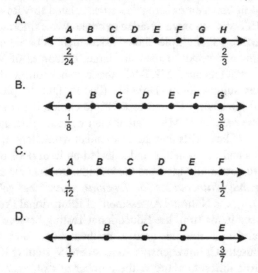

To assemble this MKT-VR assessment, we reviewed existing test banks and assessments including the Educational Testing Services (2013), Learning Mathematics for Teaching assessment (2009), NAEP (2009), and Pearson (2013a, b) to identify items that responded to the types of knowledge and skills that teachers should carry and display to demonstrate CK-VR and PCK-VR when presented with multiple-choice items. We acknowledge that the identified knowledge and skills may be a subset of all the capacities and that may be associated with MKT-VR, but given the scope of this study, we focused on those that may be most easily measured with multiple-choice items. Specifically, the knowledge and skills related to CK-VR are items that ask participants to (a) decode quantities and relationships from a VR, (b) translate a problem into a VR, and (c) find a solution from a VR. An example of a CK-VR item is presented in Fig. 5.2. The knowledge and skills related to PCK-VR are items that ask participants to identify approaches using VR to support students' specific mathematical understanding. The items related to PCK-VR include asking participants to identify approaches using VRs to support students' specific mathematical understanding, including selecting appropriate VRs for teaching specific concepts, and posing appropriate questions of VRs to promote specific understandings.

All items on the MKT-VR assessment are in the content areas of rational number, ratio, and proportion. These topics are critical content areas in middle-grades

---

ment will be used to test first hypothesis. If pilot testing and expert review indicate that we may have valid and reliable subscales measuring CK-VR and PCK-VR, we may explore treatment vs. control group differences in scores on these subscales as well.

mathematics, as emphasized by the CCSSM, and research suggests that proportionality is a unifying theme across the middle grades (Lanius and Williams 2003). The areas of rational number, ratio, and proportion are also the focus of the VAM PD project. Within these content areas, we selected items that focused on proportional relationships, number lines, and problem solving with fractions and ratios. The assessment will be given to treatment and control teachers, and in analysis, we will compare scores of treatment teachers to scores control teachers, controlling for pre-test scores.

MKT-VR Open Response Exercise   As we assembled the MKT-VR multiple-choice assessment and reviewed the qualities and utilities of VRs, we determined that our study would benefit from an additional assessment to measure other aspects of VR knowledge, particularly more performance-based skills. This measure of MKT-VR is a measure that is closely related to VAM PD and is only given to treatment teachers, to look for pre-post change (and will not be used to compare treatment and control teachers).

The Open Response Exercise will measure teachers' abilities to identify and describe quantities and relationships in a VR and to compare two solutions involving VRs, by describing the strengths and limitations of each solution and/or the reasoning behind using a specific VR for a task. In the assessment, participants need to (1) read a ratio or proportional reasoning task, (2) review two completed VRs to the ratio or proportional reasoning task, (3) identify and describe the quantities and relationships shown in the different approaches, and (4) provide advantages for the use of each type VR for teaching students about key ratio and proportional reasoning concepts. An excerpt from the Open Response Exercise is presented in the Appendix.

While we considered designing an assessment that would ask participants to create their own VRs to solve a mathematics task, we determined that we did not have the capacity to score what could be a wide range of VRs. Our focus on exploring participants' abilities to identify and describe the mathematical ideas represented in a specific set of VRs and their affordances prompts participants to engage in analysis of VRs, which may support assessment of higher levels of thinking about VRs. This assessment continues to focus on many of the same qualities of VRs as the multiple-choice assessment and emphasizes the utility of VRs while also focusing attention on comparison and evaluation.

The prompts on the Open Response Exercise focus on participants' abilities to identify and discern key quantities and relationships in the VR and to analyze provided VRs for use with a specific mathematics task. In the Open Response Exercise, we seek to understand if and how participants detail and discuss the strengths and limitations of different VR-based approaches for solving the specific problem, including attending to the quantities and relationships shown in the different approaches; alignment and relations among task, VR, and mathematics; and explanation of reasoning about the details of VR use for the mathematics task.

Analysis of teacher responses will be scored using two rubrics. First, teachers' responses on identifying and describing mathematical relationships in a VR used to solve a task will present their ability to describe how to use a VR to solve a ratio and proportional reasoning task (see Appendix). To demonstrate this ability, the participant must use (or describe how to use) quantities and relationships of the VR to solve the task and must link quantities and relationships in the VR to at least one key ratio and proportional reasoning idea. Scoring of responses will differentiate among evidence and no evidence of each component of the ability. Then, teachers' responses on describing if and how to use on VR for teaching students about specific ratio and proportional reasoning concepts will present participants' ability to describe the advantages of using one VR over another for teaching unit rate problems. To demonstrate this ability, the participant must (a) accurately identify *specific* quantities, relationships, or key ratio or proportional reasoning ideas that are more or less visible in one VR than the other and (b) justify by providing one or more reasons for a specific advantage of the tape diagram/double number line for *helping students solve or understand how to solve* unit rate problems with this type of VR. Scoring of responses will differentiate among evidence and no evidence of each component of the ability. We have used the rubrics with pilot data and refined them, coordinating with the VAM project team to make sure that we are measuring the key components and abilities of MKT-VR.

5.7 Discussion and Conclusion

As we seek to measure MKT-VR, we are aware of the multiple purposes and roles of VRs and the connections to mathematical communication, and we are interested in how increased teacher knowledge of VRs, specifically as related to alignment and strategic use of VRs, relates to instructional practices and equitable opportunities to learn: accepting that VRs can support student engagement in mathematics content, tasks, and practices, teacher knowledge likely plays a key role in student *opportunities* to learn about and use VRs. Thus not only do teachers need a robust understanding of the purposes of visually representing mathematics, the roles of VRs in instruction (Stylianou 2010), and relations between accurate and inaccurate VRs to students' problem solving (e.g., Boonen et al. 2014), they need to be able to *coordinate* these different ideas and wrestle with the complexity of their VRs and students' VRs. For example, being strategic with VR use shifts according to contexts and content, and a VR can be a tool in communication even when it may include imprecise number or line placements. To address this complexity of MKT-VR, we are developing multiple measures for understanding teacher MKT-VR, and it will be important to coordinate teachers' responses across measures to better understand the teacher knowledge for mathematics as related to VRs. This paper has laid out a theoretical framework that we anticipate using to guide and benchmark future research.

In addition, research needs to continue to seek to understand how teacher knowledge relates to instruction. We posit that teacher knowledge about VRs and related instructional practice that engages students in the doing of VRs themselves are critical to supporting all students' engagement in mathematics. VRs can be a key engagement and communication tool at the middle grades and can support rich understandings of content; researchers and the CCSSM recommend the use of visual representations in elementary and middle grades, especially number lines and tape diagrams. In addition, VRs are a tool used in mathematics classes for students who are English Learners around the work. VRs were promoted in the CCSSM as writers were influenced not only by the connections to content but by the effective use of VRs in other countries' schools, particularly Japan, China, and Singapore. Since most students in Singapore public schools are native Chinese speakers, who are taught and tested in English, Singapore students' experience and success in learning as measured on international assessments influenced our own work to support mathematics teachers of English Learners, in particular, our work to build teachers' own VR understanding and skills.

To support students who are English Learners, and specifically through opportunities to learn mathematics with VRs, we first need to understand how to measure and supports teachers' representational fluency or their MKT-VR. In this project, we are focused on ratio and proportional reasoning, and related visual representations, seeking to support teacher knowledge in this area (MKT-VR) and their knowledge of instructional practices with VRs. This project is laying the groundwork around future work to investigate whether MKT-VR has other features that should be defined related to different mathematical content, and other ways to support teacher knowledge. Future research should continue to look into how to support and measure teacher knowledge across content areas and grade levels, particularly with an eye to equity and access and promoting mathematical communication and reasoning beyond algorithmic and procedural understandings.

Appendix

Excerpt from MKT-VR Open Response Assessment

Please read the following Cookie Task.

> Andrew has a cookie recipe that uses 5 cups of sugar to make 3 batches of cookies. Draw a visual representation that can help answer the question: How many cups of sugar does Andrew need to make 19 batches of cookies?

1. Ms. Martinez drew the following diagram in response to the task.

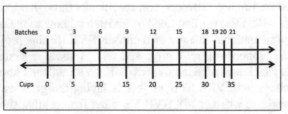

    a. List the steps you could take to solve the Cookie Task with this diagram.

    b. What are the mathematical relationships that you see in Ms. Martinez's diagram that allow you to solve the Cookie Task?

2. Mr. Diaz would like to build his sixth grade students' understanding of the mathematics in this standard, "Solve unit rate problems," using a visual representation such as a double number line or tape diagram. (Ms. Martinez's diagram is one example of a double number line. Ms. Bruno's diagram is one example of a tape diagram.) What are the possible advantages of using a double number line compared to using a tape diagram to build students' understanding of the mathematics in the above standard? Please explain as fully as you can.

References

Ball, D. L., Hill, H. H., & Bass, H. (2005). Knowing mathematics for teaching: Who knows mathematics well enough to teach third grade, and how can we decide? *American Educator, 29*, 14–46.

Ball, D. L., Thames, M. H., & Phelps, G. (2008). Content knowledge for teaching: What makes it special. *Journal of Teacher Education, 59*(5), 389–407.

Boonen, A. J., van Wesel, F., Jolles, J., & van der Schoot, M. (2014). The role of visual representation type, spatial ability, and reading comprehension in word problem solving: An item-level analysis in elementary school children. *International Journal of Educational Research, 68*, 15–26.

Campbell, P. F., Nishio, M., Smith, T. M., Clark, L. M., Conant, D., Rust, A., Neumayer DePiper, J., Jones, T., & Griffin, M. J. (2014). The relationship between teachers' mathematical content and pedagogical knowledge, teachers' perceptions, and student achievement. *Journal for Research in Mathematics Education, 45*(4), 419–459.

Chval, K., & Khisty, L. (2001, April). *Writing in mathematics with Latino students.* Presentation at the annual meeting of the American Educational Research Association, Seattle.

Common Core State Standards Initiative (CCSSI). (2010). *Common core state standards for Mathematics.* Washington, DC: National Governors Association Center for Best Practices and the Council of Chief State School Officers. http://www.corestandards.org/wp-content/uploads/Math_Standards.pdf.

DeVellis, R. F. (2012). *Scale development: Theory and applications* (3rd ed.). Thousand Oaks: SAGE Publications.

DePiper, J. N., Nikula, J., & Driscoll, M. (2014). *Professional development integrating math and language supports for students who are English Learners.* Paper presented at the National Council of Teachers of Mathematics Research Conference, New Orleans, LA.

Educational Testing Services. (2013). Praxis Series.

Gersten, R., Beckmann, S., Clarke, B., Foegen, A., Marsh, L., Star, J. R., & Witzel, B. (2009). *Assisting students struggling with mathematics: Response to intervention (RTI) for elementary and middle schools (NCEE 2009-4060).* Washington, DC: National Center for Education Evaluation and Regional Assistance, Institute of Education Sciences, U.S.

Hill, H. C., Blunk, M., Charalambous, C. Y., Lewis, J., Phelps, G. C., Sleep, L., & Ball, D. L. (2008). Mathematical knowledge for teaching and the mathematical quality of instruction: An exploratory study. *Cognition and Instruction, 26*, 430–511.

Khisty, L. L., & Chval, K. (2002). Pedagogic discourse and equity in mathematics: When teachers' talk matters. *Mathematics Education Research Journal, 14*(3), 154–168.

Lamon, S. J. (2007). Rational numbers and proportional reasoning. In F. K. Lester (Ed.), *Second handbook of research on mathematics teaching and learning* (pp. 629–667). Charlotte: Information Age Publishing.

Lanius, C., & Williams, S. (2003). Proportionality: A unifying theme for the middle grades. *Mathematics Teaching in the Middle School, 8*, 392–396.

Learning Mathematics for Teaching (LMT). (2009). *Mathematical knowledge for teaching (MKT) measures.* Ann Arbor: University of Michigan.

Lesh, R., Post, T., & Behr, M. (1988). Proportional reasoning. In J. Hiebert & M. Behr (Eds.), *Number concepts and operations in the middle grades* (pp. 93–118). Reston: Lawrence Erlbaum & National Council of Teachers of Mathematics.

Lobato, J., & Ellis, A. B. (2010). Essential understandings: Ratios, proportions, and proportional reasoning. In R. M. Zbiek (Ed.), *Essential understandings.* Reston: National Council of Teachers of Mathematics (NCTM).

Ma, L. (1999). *Knowing and teaching elementary mathematics: Teachers' understanding of fundamental mathematics in China and the United States.* Mahwah, NJ: Lawrence Erlbaum.

Massachusetts Tests for Educator Licensure (MTEL). (2011). *Middle school mathematics (47) practice test.* Amherst: Pearson Education.

Moschkovich, J. (2002). A situated and sociocultural perspective on bilingual mathematics learners. *Mathematical Thinking & Learning, 4*, 189–212.

National Governors Association Center for Best Practices, & Council of Chief State School Officers (NGA & CCSSO). (2010). Common Core State Standards: Mathematics. Retrieved from http://www.corestandards.org/assets/CCSSI_Math Standards.pdf

Ng, S. F., & Lee, K. (2009). The model method: Singapore children's tool for representing and solving algebraic word problems. *Journal for Research in Mathematics Education, 40*(3), 282–313.

Orrill, C. H., & Brown, R. E. (2012). Making sense of double number lines in professional development: Exploring teachers' understandings of proportional relationships. *Journal of Mathematics Teacher Education, 15*, 381–403.

Pearson Education, Inc. (2013a). *Mathematics teacher educator licensure exam, Mathematics Middle School* (47). Amherst.

Pearson Education, Inc. (2013b). *Mathematics teacher educator licensure exam, Mathematics Elementary* (53). Amherst.

Riley, K. R. (2010). Teachers' understanding of proportional reasoning. In P. Brosnan, D. B. Erchick, & L. Flevares (Eds.), *Proceedings of the 32nd annual meeting of the north American chapter of the International Group for the Psychology of mathematics education* (Vol. 6, pp. 1055–1061). Columbus: The Ohio State University.

Shulman, L. S. (1986). Those who understand: Knowledge growth in teaching. *Educational Researcher, 15*, 4–14.

Siegler, R., Carpenter, T., Fennell, F., Geary, D., Lewis, J., Okamoto, Y., Thompson, L., & Wray, J. (2010). *Developing effective fractions instruction for kindergarten through 8th grade: A practice guide (NCEE #2010-4039)*. Washington, DC: National Center for Education Evaluation and Regional Assistance, Institute of Education Sciences, U.S. Department of Education Retrieved from http://ies.ed.gov/ncee/wwc/pdf/practice_guides/fractions_pg_093010.pdf.

Siegler, R. S., Thompson, C. A., & Schneider, M. (2011). An integrated theory of whole number and fractions development. *Cognitive Psychology, 62*(62), 273–296.

Singh, P. (2000). Understanding the concepts of proportion and ratio constructed by two grade six students. *Educational Studies in Mathematics, 43*(3), 271–292.

Smith, J., & Thompson, P. W. (2007). Quantitative reasoning and the development of algebraic reasoning. In J. Kaput, D. Carraher, & M. Blanton (Eds.), *Algebra in the early grades* (pp. 95–132). New York: Erlbaum.

Stylianou, D. A. (2002). On the interaction of visualization and analysis: The negotiation of a visual representation in expert problem solving. *The Journal of Mathematical Behavior, 21*(3), 303–317.

Stylianou, D. A. (2010). Teachers' conceptions of representation in middle school mathematics. *Journal of Math Teacher Education, 13*, 325–343.

Stylianou, D. A. (2011). An examination of middle school students' representation practices in mathematical problem solving through the lens of expert work: towards an organizing scheme. *Educational Studies in Mathematics, 76*, 265–280.

Stylianou, D. A., & Silver, E. A. (2004). The role of visual representations in advanced mathematical problem solving: An examination of expert-novice similarities and differences. *Mathematical Thinking and Learning, 6*(4), 353–387.

U.S. Department of Education, Institute of Education Sciences, National Center for Education Statistics, National Assessment of Educational Progress (NAEP). (2009) Mathematics Assessment. Retrieved from http://nces.ed.gov/NationsReportCard/nqt/DownloadOutput/DownloadOutput/11860.

Woodward, J., Beckmann, S., Driscoll, M., Franke, M., Herzig, P., Jitendra, A., Koedinger, K. R., & Ogbuehi, P. (2012). *Improving mathematical problem solving in grades 4 through 8: A practice guide (NCEE 2012-4055)*. Washington, DC: National Center for Education Evaluation and Regional Assistance, Institute of Education Sciences, U.S. Department of Education Retrieved from http://ies.ed.gov/ncee/wwc/publications_reviews.aspx#pubsearch/.

Zbiek, R. M., Heid, M. K., & Blume, G. (2007). Research on technology in mathematics education: The perspective of constructs. In F. K. Lester (Ed.), *Second handbook of research on mathematics teaching and learning* (pp. 1169–1208). National Council of Teachers of Mathematics: Reston, VA.

Part II
PCK in Formal Pre-service Teacher Learning

Chapter 6
Teacher Inquiry as a Vehicle for Developing Pedagogical Content Knowledge in Pre-service Teachers

Marisa Harford, Rachel Leopold, Wendy Thomas Williams, and Elizabeth Chatham

Abstract This case study describes the interactions between pedagogical content knowledge (PCK) and practitioner inquiry as observed in a residency-model teacher preparation program. In the Math and Science Teacher Education Residency (MASTER), novice teachers engaged in iterative cycles of assessing student thinking and using that assessment to reflect on and inform instructional decisions. Through an analysis of resident presentations on their inquiry and examination of student work, this case study explores the synergy between inquiry processes and PCK development and demonstrates that using teacher inquiry as a vehicle for building and refining PCK is a promising practice. MASTER pre-service teachers utilized canonical PCK to analyze assessments, develop goals for student learning, plan for instruction, and then interpret the results of further assessment to understand the impact of their instructional strategies and how they might revise their strategies for future lessons. Implementing the inquiry process provided opportunities for residents to build their personal PCK through their reflection on and analysis of their students' learning and their own teaching. Collaboration protocols allowed coaches and peers to provide prompts that engaged directly with the content the resident was teaching, pushed the resident to deepen thinking, and developed shared understandings through integrating PCK from multiple individuals. The case study investigates the ways in which PCK is foundational to conducting teacher inquiry but is also built and refined by testing through the "inaction" reflection and revision fostered by the teacher inquiry process.

Keywords PCK · Teacher inquiry · Pre-service learning · Teacher residency · Inquiry cycle · Canonical PCK · Personal PCK · Student thinking

---

M. Harford (✉) · W. T. Williams · E. Chatham
New Visions for Public Schools, New York, NY, USA

R. Leopold
New York City Department of Education, New York, NY, USA

6.1   Introduction

In 2013, three partners – New Visions for Public Schools; Hunter College, CUNY; and the New York Hall of Science – launched a new teacher residency, the Math and Science Teacher Education Residency (MASTER) program. The MASTER program sought to improve student learning in math and science in New York City public schools through recruiting, preparing, and retaining a cadre of high-quality novice teachers ("residents") and through simultaneous development of experienced math and science teacher mentors who served as these residents' coaches and guides throughout a year-long intensive clinical experience that provided a context for applying and practicing theories and strategies learned within the coursework for a Master's degree in secondary education from Hunter College.

Becoming an effective teacher requires understanding, applying, and continuously developing a complex set of interconnected concepts, including knowledge of content, students, educational goals, curriculum, and instructional strategies, and how to bridge these areas based on specific student learning needs and contextual factors. The partners understood that in order to develop effective math and science teachers, they needed to attend carefully to the construct of pedagogical content knowledge (PCK), that is, "...for the most regularly taught topics in one's subject area, the most useful forms of representation of those ideas, the most powerful analogies, illustrations, examples, explanations, and demonstrations - in a word, the ways of representing and formulating the subject that make it comprehensible to others" (Shulman 1986), and ensure that residents experienced learning activities during their preparation year that would foster the development of PCK. This paper examines one of the central approaches used by the MASTER program in order to develop residents' PCK: teacher inquiry. Through an in-depth case study of three science residents' artifacts of practice, it explores whether and how inquiry-based learning activities supported these novice teachers in building aspects of their PCK. In the MASTER program, inquiry is defined as the iterative process of assessing student thinking and using that assessment to reflect on and inform instructional decisions such as planning what to teach, how to teach it, and how to differentiate or adjust instruction for different groups or individuals. This study explores the following research questions related to novice teacher inquiry and PCK:

- In what ways does engaging in the inquiry process help novice teachers to develop PCK?
- What evidence is present in novice teachers' analysis of student work and reflections on teaching that they have developed or are making use of PCK? Are there patterns in the type of evidence that are present, or the frequency of the appearance of certain types of PCK-related capacities?
- What types of structures or prompts (e.g., questions, protocols) develop novice teacher capacity in inquiry and PCK?

6.2   Review of Literature

Since Shulman's coining of the term pedagogical content knowledge in 1986, and the advent of the "teacher as researcher" movement in the 1970s, many researchers have contributed to our growing understanding of both teacher inquiry (or action research) and PCK and the role these two constructs play in teacher decision-making and development. Rather than describing the full spectrum of the literature on PCK and teacher inquiry, this review will focus on a few key strands of thought that are most relevant to this case study.

Central to both inquiry and PCK is the view of teachers as meaning-makers and the belief that teacher decision-making and reflection represent crucial components of the act of effective teaching (Duckworth 1986). Theories of PCK emphasize that teachers must bridge knowledge of content, students, and pedagogy in order to implement any lesson (Gess-Newsome and Lederman 1999, Gudmundsdottir 1987). Shulman (1986) described the development of PCK as taking place through "comprehension, transformation, instruction, evaluation, and reflection" but noted that these elements would be enacted in different sequences and combinations. While the foundations of PCK can be learned and shared by professors, mentors, and others, in the moment of the implementation of a lesson, the teacher must rely on his or her own PCK to guide the hundreds of small decisions made in the course of a class period – how to interpret and respond to student comments and ideas, how to check for understanding, how to address misconceptions, and how to adjust instruction "inaction" based on these factors (Magnusson et al. 1999). Similarly, in an inquiry model, teachers act as researchers in their own classrooms, generating new knowledge about teaching through deliberate study of the impacts of their instructional choices on student learning or other outcomes (Langer et al. 2000). Akin to the application of PCK, this knowledge is contextual and specific to the interaction of a group of students with a body of content in specific circumstances but can also be generalized to inform the teacher's (and others') work in the future.

Much of the recent research on PCK has focused on the multifaceted nature of the construct and described how those facets interact with and impact each other. Ball, Thames, and Phelps (2008) proposed that Shulman's content knowledge could be subdivided into common content knowledge (CCK) and specialized content knowledge (SCK), and his pedagogical content knowledge could be divided into knowledge of content and students (KCS) and knowledge of content and teaching (KCT). This approach emphasized the importance of teachers anticipating what students are likely to think and what they will find confusing to inform decisions about how to sequence particular content, which examples to use, and which models and analogies are most helpful for teaching particular content. Park, S. and Oliver, J. S. (2008) hypothesized that the development of one component of PCK may simultaneously encourage the development of others and ultimately enhance the teacher's overall PCK; the components are ultimately more meaningful when interrelated. Their data analysis revealed that PCK development occurred as a result of reflection related to both "knowledge-in-action" and "knowledge-on-action" and that students influenced the ways that PCK was organized, developed, and validated. The interrelationship between knowledge-in-action and knowledge-on-action implies that PCK development encompasses knowledge acquisition and knowledge use and that it is unlikely that teachers acquire PCK first and then enact it. Rather, teachers develop PCK through the dynamics of knowledge acquisition, new applications of that knowledge, and reflection on the uses embedded in practice.

Especially relevant to our case study is the synergistic relationship between canonical and personal PCK posited by Smith, Plumley, and Hayes (2016). They describe a distinction between canonical PCK, which is documented in the literature and learned from external sources, and personal PCK resulting from teachers' lived experiences and reflections on classroom practice. They note that the application of canonical PCK in real settings leads to reality testing, adding to, revising, and adjusting canonical PCK in light of one's personal teaching experience. In addition, a teacher's knowledge and proficiency in assessment and use of assessment to continually monitor student outcomes and make teaching decisions also inform the teacher's PCK (Morine-Dershimer and Kent 1999). PCK also helps teachers to navigate the process of aligning their teaching to research-based best practices as well; for example, deeper PCK is correlated with science teaching that is more consistent with the principles of scientific inquiry (Park et al. 2011).

A similar focus on the relationship between received or canonical knowledge about teaching and the process of teacher growth through implementation, assessment, and reflection lies at the heart of the "teacher as researcher" or inquiry approach to educator development, which emphasizes the teacher's dynamic, ongoing learning. Cochran-Smith and Lytle (1993) define teacher research as "systematic, intentional inquiry by teachers about their own school and classroom work." They claim that "only teachers themselves can integrate their assumptions and their interpretive frameworks and then decide on the actions that are appropriate for their local contexts." Inquiry, as operationalized in the MASTER program, fits within the definition of teacher action research.

In order to prepare novice teachers for effectiveness in New York City, the designers of the MASTER program selected a model of teacher inquiry aligned to the school-based collaborative inquiry work in use across the district since 2008. For our purposes, we define teacher inquiry as the pursuit of knowledge and solutions through the "systematic, intentional" study of practice (Cochran-Smith and Lytle 1993) with the goals of developing resident PCK and meeting student outcomes. The core features of this model include the use of structured protocols; identification of a focus, criteria for success, and a specific goal based on initial data-gathering; grounding the inquiry in analysis of student work and evidence of student thinking (interviews, assessments) in order to understand the impact of teacher instructional strategies and assess progress; implementation of iterative cycles to refine and develop the work; final reflection and incorporation of changes into practice; and collaboration or peer feedback.

In their research on inquiry models for school improvement in New York City, Talbert et al. (2012) note that inquiry fosters shifts in teacher focus from teaching to student learning, from summative to formative assessment, and from external attributions of student failure to instructional efficacy. Teacher inquiry places student learning, rather than teacher actions, at the center of educator development and learning, and allows educators to take ownership over student data and related decisions (Huffman and Kalnin 2003, Robinson 2010). Teachers build their practice through an emphasis on the use of data to inform evidence-based practice and "just-in-time" content-specific professional development – that is, teachers learn new instructional strategies when they need them to help students reach a learning goal and integrate their lived experience in the classroom with systematic reflection (Dawson, 2014). Analysis of student work as a systematic approach to professional inquiry helps determine the teaching methods that are working successfully (Langer et al. 2003) and supports novice teachers in being open to feedback and improving their instruction (Elliott, 1988). Further, Langer, Colton, and Goff have found that through purposeful analysis of student learning, over time, transformative teacher learning occurs. The same holds true in inquiry processes for novice teachers. As documented by Tabachnick and Zeichner (1999), "More quickly than is usually the case for prospective teachers, they [new teachers] shifted their focus away from themselves as teachers to their students as learners. The process of doing action research, including as it does the gathering of data about student learning, encouraged this shift in focus."

Because teacher inquiry centers around evidence of student learning, within the context of teacher preparation, it can be used for the dual purpose of supporting student outcomes and building educator capacity. In a study on teacher inquiry in professional development schools, Silva and Dana (2001) found that developing a collaborative inquiry stance simultaneously improves outcomes for students and preparation for novice teachers. Teacher inquiry strategies have demonstrated impact on student learning in the classrooms of in-service teachers as well (Grace et al. 2015, Zargarpour 2005). Saunders, Goldenberg, and Gallimore (2009) found significant achievement gains were produced for students when grade-level teams used a structured and consistent collaborative inquiry process. From 2009 to 2014, two of the MASTER program partners, New Visions for Public Schools and Hunter College, implemented the Urban Teacher Residency (UTR), utilizing the same inquiry model used in MASTER in the preparation of novice teachers. In their summative evaluation of UTR, external evaluators Rockman et al. (2014) found that students taught by UTR-trained teachers—both residents and graduates—consistently outscored peers on Regents exams in math and science.

In requiring novice teachers to implement and reflect on an inquiry process during their pre-service year, the MASTER program hoped to build both PCK and lifelong instructional decision-making skills in its residents. Ideally, residents become the type of teachers that Sagor (2009) envisioned as utilizing certain "habits of mind" when they are confronted with challenges with student learning – including assessment, implementing informed interventions, monitoring and revising those interventions, and reflecting on successes and challenges to generalize their learning. The design of the MASTER program incorporated ongoing development of PCK through engagement in inquiry processes, culminating in a capstone inquiry project, the Defense of Learning, which is grounded in these principles and predicated upon the notion that effective teaching requires meaning-making and that it is the role of a teacher education program to foster those capacities.

6.3 Conceptual Framework: Operationalizing PCK and Inquiry and How They Impact Each Other

The MASTER program partners collaboratively developed a set of PCK learning objectives to guide the program's focus on novice teacher development. These learning objectives, based on the PCK literature and refined over the course of the project, guided the design of performance-based assessments of resident practice and aligned clinical and coursework experiences.

6.3.1 PCK Learning Objectives

Mathematics and Science teachers with a high level of PCK integrate knowledge of their content with an understanding of how to communicate that knowledge to students in a way that fosters deep understanding through:

1. Building on their own conceptual models of mathematical and scientific phenomena to support their students in developing conceptual understandings.
2. Knowing the key concepts and practices secondary students must learn in their content and articulating how students might progress through stages or phases of understanding over time.
3. Understanding how students learn in their content area and anticipating the prior knowledge, initial thinking or misconceptions, and skills students will bring to a task.
4. Employing multiple instructional models and approaches to help students access and understand content, concepts, and practices.
5. Making instructional decisions that are based on student thinking and addressing misconceptions or disciplinary skill gaps.
6. Using formative assessment to continually evaluate and monitor student understanding of content, concepts, and practices.

For the purposes of this study, we focused on objectives #5 and #6, which were most easily observed in resident work products related to inquiry.

6.3.2   The Inquiry Cycle

During the residency year, MASTER residents engaged in iterative cycles of teacher inquiry that culminated each semester in a "Defense of Learning" presentation. Given our view of teacher inquiry as the pursuit of knowledge and solutions through the "systematic, intentional" study of practice (Cochran-Smith and Lytle 1993), residents were prepared to become reflective practitioners seeking to continuously improve their pedagogical skills. Through monthly inquiry seminars and ongoing work guided by mentor teachers, program coaches, Hunter College professors, and peers, residents used the following inquiry process to inform their instructional decision-making and meet student learning goals:

1. Use a diagnostic/baseline assessment to select a focus skill or content goal, and describe students' initial understanding of the focus skill or content.
2. Establish a learning goal(s) for students and an initial instructional plan.
3. Implement the instructional plan.
4. Assess and analyze student learning through formative assessment, using specific protocols for looking at student work.
5. Revise instructional strategies based on the analysis, and implement the revised strategies.
6. Complete at least three cycles of steps #3–5.
7. Administer a summative assessment to capture evidence of student progress.
8. Reflect on implications of the assessment results and the inquiry cycle; present the inquiry process to mentor, program coach, administrators, and peers, using student work and data as evidence (Fig. 6.1).

The monthly seminar sessions outlined in the seminar syllabus guided residents through each phase of the inquiry process and provided opportunities for them to receive feedback on their thinking and analysis from peers and program coaches. Structured protocols related to task analysis, for collaboratively examining student work, and using assessment to drive instruction (the keep-start-change instructional strategies protocol), as well as the rubric for the Defense of Learning presentation itself, were used to guide residents' thinking around how to apply their growing PCK to improve their students' learning and build their own teacher decision-making capacities.

Drawing from these learning objectives and research around teacher inquiry, we developed a conceptual map to explicate the relationships we hypothesized between the inquiry cycle and the growth of novice teacher PCK. In diagramming these interactions, we noted that PCK is essential for all of the activities of the inquiry cycle, and in turn, the results of the cycles of inquiry should result in further refinement and development of the resident's personal PCK (Fig. 6.2).

Through the inquiry process, teachers are asked to apply and build on the following aspects of their PCK:

- Knowledge of the key concepts and practices in the curriculum (e.g., state standards) and how students might progress through them in order to set measurable goals and specific instructional objectives that align to the curriculum and take into account student initial understandings.

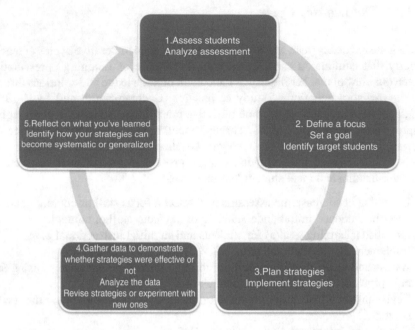

Fig. 6.1 The MASTER teacher inquiry cycle

- Strong conceptual understanding of the material, the common ways students' understanding might progress, and how those stages might be made visible in student work.
- "Toolbox" of content-specific instructional strategies and the ability to articulate how each strategy might support student learning of a particular concept, practice, or skill; the ability to select an appropriate instructional strategy based on content, context, and assessment information about student understandings
- Ability to analyze student work and responses to notice content- and discipline-specific patterns of thinking and ideas that students may demonstrate in their work and draw connections between instructional strategies used and results in terms of student thinking.
- Ability to integrate specific observations from student work with generalizations, patterns, and trends from canonical or received sources of PCK.

6.4   Methods

6.4.1   Data Collection and Sample

In order to understand teachers' PCK, it is also necessary to move beyond their "declared PCK" to examine the teaching decisions they make for evidence of PCK (Thompson, 2016). This case study examined transcripts of oral presentations and

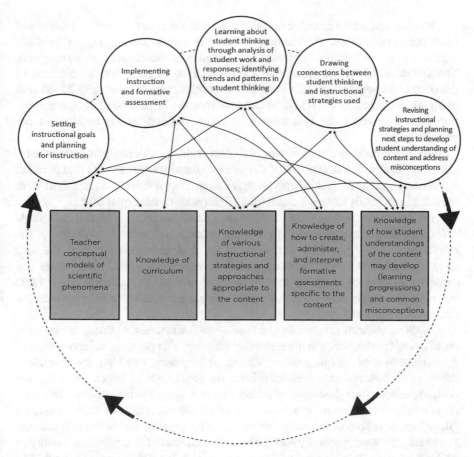

Fig. 6.2 Interactions between PCK and teacher inquiry processes. (Graphic design: Kerrianne Eames)

discussions of analysis of student work from three science teacher residents who took part in the MASTER program during the 2015–2016 school year, with additional data included from written work samples from a task analysis activity produced by those three residents and other science residents from the same cohort.

The Looking at Student Work discussions took place in the context of the monthly inquiry seminars facilitated by the MASTER program coaches. These collaborative conversations were characterized by significant give-and-take between the presenting resident, peers, and program coach who was facilitating the Looking at Student Work activity. Each session lasted approximately 10 min. They were video recorded, and transcripts were made of the videos to facilitate coding and analysis and provide confidentiality.

The case study also examines transcripts from the three residents' Defense of Learning presentations, the culmination of the semester-long inquiry process. Again, the presentations were video recorded and the recordings transcribed. The Defense of Learning is a formal presentation, so there was no element of question and answer involved. Given that the presentations are usually 25–35 min in length, we selected a 10–12-min excerpt from the beginning of each presentation that represented the first inquiry cycle as defined by the resident in the presentation.

Finally, the residents' written work samples from the task analysis activity were also completed within the context of the monthly inquiry seminar – the first iteration in February and the second in May. Samples from other residents in the cohort were included in the analysis so we could identify broader trends or patterns.

In selecting the three residents for the case study, we focused on science to narrow down the content area and align to the professional areas of expertise of the researchers coding the data. Three out of the four individuals coding the data have experience as science educators in K-12 classrooms, and two of those had extensive experience coaching science teachers as well; the remaining individual has more than 6 years of experience coaching novice science teachers within the context of a residency model.

The three residents selected for the case study deliberately represent various levels of overall performance in the residency program. The program utilizes a suite of performance-based assessments, including observations rated on the Danielson rubric as well as program-specific rubrics for areas such as lesson planning and professionalism. Residents are required to meet program benchmarks but may require revisions or multiple attempts to do so. One of the case study residents, "Ilana," exceeded program benchmarks on all of her assessments on the first attempt. A second, "Steven," generally met program benchmarks throughout the residency on his first attempt, with only one revision needed, but did not exceed them. The third, "Beth," struggled to meet program benchmarks and required revisions to meet standards in several areas. Beth was preparing to become a teacher of high school biology, Ilana of middle school integrated science, and Steven of high school chemistry. We selected two female residents and one male resident, which reflected the gender composition of the cohort.

6.4.2 Data Analysis

Initially, we used relevant literature and foundational program documents to develop the conceptual framework and descriptive vocabulary to categorize patterns and trends observable in residents' work. We then used the conceptual framework to create an initial set of categories for coding the data. After developing the initial codebook, all four authors collaboratively normed around the codes using the Defense of Learning transcripts, clarifying and articulating more specific criteria for applying the codes and noting key examples for each code. Based on this norming and refinement of the code book, the authors then completed some coding of additional transcripts individually. Due to the limited data set, the group was able to come back together and collaboratively review the coding for each of the Defense of Learning transcripts so that we reached consensus around the codes that were applied. We also collaboratively reviewed one of the Looking at Student Work transcripts together to reach consensus on our coding criteria given that the parameters and context of those transcripts were different from the Defense of Learning samples, and then a subset of the authors coded the remaining transcripts.

When we began to work on the Looking at Student Work transcripts, we also incorporated a parallel set of codes related to prompts from program coaches or peers that were designed to elicit responses in one of the applicable categories. For example, in Steven's Looking at Student Work session, the coach prompted him to describe patterns and trends in the student data by asking, "What was the proportion of students that you feel like got it versus didn't?" That type of question was coded as a prompt that supported the resident's inquiry process. Observations about data could also be coded as prompts, such as the comment that a resident's peer made about a sample of Beth's students' work: "It [the student's work] says if they're reproducing, they will mutate. It makes me think that the student is confusing sexual reproduction and the variation that's produced through crossing over and mutation." This type of contribution to the analysis was also coded with a parallel code that indicated it represented an analysis of student thinking, but not by the presenting resident.

6.4.3   Coding Schema

Six codes were ultimately identified to apply to the transcript and work sample data, and the team aligned them to the two focus PCK objectives and the relevant steps of the inquiry process. Given the focus of this analysis, one criterion common to all codes was a reference to discipline-specific content or practices. An instance where general instructional strategies were referenced – for example, an extended discussion that occurred during Beth's Looking at Student Work discussion about how long to allow students to complete their summary activity at the end of class – was not coded unless it focused on students' learning of specific science content or practices. These might be valuable discussions for the residents in terms of the development of their teaching, but they did not represent using or building pedagogical content knowledge (Table 6.1).

Table 6.1 Coding schema

PCK objective	Step of the inquiry process	Evidence	Elements (requires majority of elements)	Examples from the transcripts of excerpts classified with this code
A. Using formative assessment to continually evaluate and monitor student understanding of content, concepts, and practices	A1. Planning and implementing appropriate assessments of progress and methods of gathering data about the effectiveness of instructional strategies	Description of how student progress toward objectives will be measured and how that will connect to evaluating the effectiveness of teaching strategies	1. Assessment method, including reference to specific content or disciplinary practice being assessed 2. Assessment criteria, including reference to specific content or disciplinary practice being assessed 3. Connection to instructional strategy	"I used the CER framework... within that, a successful explanation would have a claim written as a statement..." "the green box [is] where I used to assess whether or not they were strong in making a strong conclusion at this point with constructing information. So here they are synthesizing what they gathered from their model and explaining why there's certain relationships between energy and bonds" "What student B did well was incorporate an accurate claim, for providing some supporting data and observations from the model... But again, this student was also missing the underlying chemistry concepts and principles to explain why that thing is true" "in question 6, he did mention the word common ancestor which was exactly what I was looking for. He said that species B became extinct and was a common ancestor for C and D... Based on that, he seems to have a better understanding of common ancestors and relationships showing on evolutionary trees"

A2. Analyzing individual students' work and responses to make observations and reach valid inferences about students' understandings, knowledge, skill, and use of disciplinary practices	Shares an observation or noticing from student work and makes an inference about what that means about the student's understanding of the specific content or disciplinary practice	1. Specific observation from student work/response 2. Inference about student's strengths, challenges, understandings, *or* misconceptions based on the observation 3. Reference to specific content or disciplinary practice	"Based on what all of them can do, they can all calculate the molar mass of a substance which was from a previous lesson" "we found that on the January Mock Regents, a huge majority of both classes combined, 50 out of 58 students, answered this following question incorrectly. It was about the reactions and the energy and the bonds involved in reactions. So a lot of students take the red answer, where energy is absorbed as the bond is formed. When in this reaction bonds are forming, but as bonds are forming, energy is actually released"
A3. Identifying patterns and trends in groups of students' understanding, knowledge, skill, and use of disciplinary practices	Describes a pattern, trend, or group of similar students' performance along with an inference about what that means about the students' understanding of the specific content or disciplinary practice	1. Inference about students' strengths, challenges, understandings, *or* misconceptions based on the observation 2. Description characterizing or grouping according to a pattern/trend 3. Reference to specific content or disciplinary practice	"It requires energy to break a bond which is what I wanted students to see with the magnets..I wanted them to use that model to make a conclusion and cite evidence from that model to help them explain what's going on with bonds and energy in a reaction" "So this was the CER graphic organizer that they had for the first lab, which was inertia around a curve... they took a marble and went around a track. Then they observed how it left an opening… This was the framework that I gave them. So [I gave them] a table to sort out each of the different components and then at the end, they had to put it back together in paragraph form"

(continued)

Table 6.1 (continued)

PCK objective	Step of the inquiry process	Evidence	Elements (requires majority of elements)	Examples from the transcripts of excerpts classified with this code
B. Making instructional decisions that are based on student thinking and addressing misconceptions or disciplinary skill gaps	B1. Planning and implementing instructional strategies designed to support students in reaching the instructional objective that takes student thinking into account	Description of a specific teaching strategy with a connection to why it is an appropriate strategy for helping students to learn specific content or disciplinary practices	1. Specific teaching strategy 2. Rationale for why this strategy was selected 3. Reference to specific content or disciplinary practice	"Within reasoning, 70% of students were able to provide the scientific concept that was related but weren't able to connect it with their claims and evidence. So a lot of the times they would just say, oh it's inertia, and then define what inertia was. Their reasoning may have been then, I have provided the concept, which is what was asked of them in the rubric and in what we've gone over. But they weren't making those connections clear for the rest of what they were observing and what that tie to the concepts were" "I didn't realize students… probably don't know what you mean when you say explain… so I think next time I'll give them that sheet that they provided, with the different terms at the beginning…"
	B2. Drawing connections between instructional strategies used and the results in terms of student thinking	Description of how a teaching strategy impacted how students thought about, understood, or applied the content or disciplinary practice	1. Specific teaching strategy 2. Evidence from student work/responses (individual or pattern) 3. Inference about impact of teaching strategy based on evidence from student work 4. Rationale/hypothesis about why the teaching strategy had this impact 5. Reference to specific content or disciplinary practice	"So next steps, based on this, [were] to keep using the graphic organizer with the prompts… emphasizing more of what's expected within the reasoning section and how should that look. And then, changing to use tests that result in quantitative data so things that were really numbers based might be a little more concrete for them" "The majority of students had a strong understanding what a redox reaction is. I think it was more of the content part of it, where they couldn't identify the oxidation number to begin with… Definitely provide them kind of a more step by step visualization, and the go-to rules because it's really a dry kind of process to identify oxidation numbers…"

B3. Revising instructional strategies based on assessment evidence	Description of a change in teaching strategy based on an inference about how the use of that strategy impacted how students thought about, understood, or applied the content or disciplinary practice	1. Revision to teaching strategy 2. Rationale for change based on how the prior strategy impacted student learning 3. Reference to specific content or disciplinary practice

6.5 Results and Discussion

6.5.1 Coding Tallies (Tables 6.2, 6.3 and 6.4)

6.5.2 Observations from the Data: In What Ways Does Engaging in the Inquiry Process Help Novice Teachers to Develop PCK?

6.5.2.1 Evidence of Use of PCK During the Inquiry Cycle

Based on these three residents' summative Defense of Learning presentations, it is clear that these novice teachers utilized a great deal of PCK in the aspects of the inquiry cycle examined by this case study, including planning for and implementing

Table 6.2 Coding tally for Defense of Learning excerpts

PCK objective	Using formative assessment to continually evaluate and monitor student understanding of content, concepts, and practices				Making instructional decisions that are based on student thinking and addressing misconceptions or disciplinary skill gaps			
Resident	A1 Assessment	A2 Individual analysis	A3 Trend analysis	Totals	B1 Planning for student thinking	B2 Strategies and student thinking	B3 Revising instructional strategies	Totals
Ilana	2	1	5	8	4	2	1	7
Steven	3	4	11	18	2	1	1	4
Beth	0	3	1	4	1	0	0	1
Totals	5	8	17	30	7	3	2	12
Average	1.7	2.7	5.7	10	2.3	1	0.7	4

Table 6.3 Coding tally for Looking at Student Work discussions

PCK objective	Using formative assessment to continually evaluate and monitor student understanding of content, concepts, and practices				Making instructional decisions that are based on student thinking and addressing misconceptions or disciplinary skill gaps			
Resident	A1 Assessment	A2 Individual analysis	A3 Trend analysis	Totals	B1 Planning for student thinking	B2 Strategies and student thinking	B3 Revising instructional strategies	Totals
Ilana	3	0	7	10	3	2	1	6
Steven	3	1	1	5	1	0	1	2
Beth	1	1	2	4	1	0	1	2
Totals	7	2	10	19	5	2	3	10
Average	2.3	0.7	3.3	6.3	1.7	0.7	1.0	3.3

Table 6.4   Coding tally for Looking at Student Work – prompts by Coach or Peer

PCK objective	Prompts related to using formative assessment to continually evaluate and monitor student understanding of content, concepts, and practices				Prompts related to making instructional decisions that are based on student thinking and addressing misconceptions or disciplinary skill gaps			
Resident	PA1 Assessment	PA2 Individual analysis	PA3 Trend analysis	Totals	PB1 Planning for student thinking	PB2 Strategies and student thinking	PB3 Revising instructional strategies	Totals
Ilana	3	0	5	8	3	3	1	7
Steven	2	4	5	11	1	0	1	2
Beth	1	5	5	11	3	2	1	6
Totals	6	9	15	30	7	5	3	15
Average	2.0	3.0	5.0	10.0	2.3	1.7	1.0	5.0

assessments, analyzing individual students' work, identifying trends and patterns in the work of a group of students, planning and implementing instructional strategies that take student thinking into account, drawing connections between instructional strategies used and the results in terms of student thinking, and revising teaching strategies based on inferences about how the use of that strategy impacted student thinking. With the exception of a section in Beth's analysis of student work session (focused on how much time students should be given to complete an exit ticket), the vast majority of the conversations and presentations analyzed made explicit reference to disciplinary content and practices and were focused on how to interpret student thinking and apply instructional strategies related to the specific content that residents were teaching.

There were also examples in the transcripts that revealed the residents utilizing their developing PCK in order to make sense of the content and place it in the context of the "big ideas" of their curriculum. For example, in Ilana's Looking at Student Work session, she noted, "I've been trying to use this common umbrella of changes over time within the evolution unit, and…mutations and passage of genetic material are changes over time." Here, the process of going through the inquiry cycle – especially analyzing student work to uncover the concepts that students do and do not understand at any given point – requires that the resident reference their PCK about the scope of the curriculum and the crosscutting concepts. Ilana was able to notice a trend in student thinking based on specific examples of work and connect it to what she knew about the curriculum overall in order to inform her decision-making about what to address next in her teaching of these specific students.

6.5.2.2   Interactions Between Canonical and Personal PCK

Several moments in the transcripts pointed explicitly to the interaction of canonical PCK with personal PCK. For example, in her analysis of student work session, Beth noted that, based on her work, a student seemed to think that organisms can adapt individually. One of her peers replied, "Individually adapting. That's a really common misconception…In our evening biology class, [the professor] made the point that college and grad school level people still have that misconception. It's really hard to get rid of it." In this example, the peer characterized a specific student's idea as an example of a canonical misconception they had learned about in their graduate school work, connecting this particular sample of student work to an idea from the PCK literature.

In general, in their analysis of student work, residents were able to characterize student responses in terms of students' thinking and understanding of the content, not just as correct or incorrect. Making a reasonable inference about what a particular student's response indicates in terms of their thinking about the content requires PCK related to how students might express their ideas and possible naive conceptions of the content. This excerpt from Steven's Defense of Learning provides a clear example of how he starts with quantitative data about outcomes on an assessment but quickly moves to describing what the implications are for his understanding of students' strengths and weaknesses in terms of content and science practices such as using evidence in scientific writing:

> …what I found was that 78 percent of students… in both chemistry classes, they were scoring between the 1 and 2 range [on the rubric]. Basically what we found is they were making pretty accurate claims in their conclusions. But what they weren't doing was explaining the underlying scientific concepts and they weren't using data to support their conclusions, the data from their lab and the data they collected. Another piece of data, on the content end, was we found that on the January Mock Regents, a huge majority of both classes combined, 50 out of 58 students, answered this following question incorrectly [shows question on powerpoint]. It was about the reactions and the energy and the bonds involved in reactions. So a lot of students take the red answer, where energy is absorbed as the bond is formed. When in this reaction bonds are forming, but as bonds are forming energy is actually released…[the students] were having a conceptual gap in the nature of bonding and in reactions. The misconception…was that in a reaction when bonds between atoms are broken, energy is always being released…

During the Looking at Student Work section, peers and coaches also provided prompts to build residents' skills in this area. For example, in Steven's session, a peer noted the following about the student work sample they were reviewing, "… that student mostly didn't memorize what the three different types of reactions are, but has the important concepts about what's happening with the electrons." Here, the peer's comment functions to summarize a small segment of the discussion and focus the conversation around what the student does and does not understand about the content. PCK is essential to this ability to make an inference about student thinking based on a sample of student work.

6.5.3   What Evidence Is Present in Novice Teachers' Analysis of Student Work and Reflections on Teaching that They Have Developed or Are Making Use of PCK?

6.5.3.1   Identifying Patterns in Students' Understanding of Disciplinary Practices and Concepts

Another trend in the data was the frequency with which residents examined students' use of scientific practices to learn and demonstrate their understanding of the content, coded as A3 – identifying patterns and trends in groups of students' understanding, knowledge, skill, and use of disciplinary practices. Both Ilana and Steven focused their Defense of Learning presentations on how students communicated their grasp of core concepts in the science curriculum through writing, especially around the use of scientific evidence and reasoning to support their ideas. In Ilana's Defense of Learning, she introduced her instructional goals this way:

> In writing scientific explanations…I used the framework of claim-evidence-reasoning scientific explanations, or CER. So from within that, a successful explanation would have a claim written as a statement that answers the question being asked [and] use specific evidence that supports the claim. Reasoning includes scientific concepts that accurately connect the evidence to the claim. So they can state what the concept is and then be able to tie it in to why it's explained in the evidence itself, and also using accurate academic vocabulary…, in taking this criteria, I applied it to this baseline for balanced and unbalanced forces with a small activity…We went over their analysis questions, looking at different scenarios, they had to state whether it [the force] was balanced or unbalanced, so that would be their claim. And also, what was your evidence and your reasoning for your decisions.

Here, Ilana shows how she is connecting students' use of science practices such as communicating using evidence and scientific reasoning to students' learning of the concept of balanced and unbalanced forces. In order to develop and analyze the assessment she describes, she utilized PCK that included knowledge of the science curriculum and standards, content knowledge, knowledge of science practices, knowledge around how to use writing to assess student understanding of that content knowledge, and knowledge of how to communicate criteria for effective scientific writing to students. All of these various elements of PCK were required to implement this one initial step of the inquiry process (coded as A3) which is an enactment of our PCK objective A, "Using formative assessment to continually evaluate and monitor student understanding of content, concepts, and practices."

In addition, our analysis surfaced examples of residents explaining how their planning connected to fostering student development of science concepts. In her Looking at Student Work session, Beth described a simulation activity she used with her students to build their conceptual understanding:

> I wanted them to explain how bacteria builds resistance through mutation…natural selection will occur, and then…the favorable trait will only be present in the later generations. And I wanted them to tie that in with real life and explain why doctors tell us to finish the entire course of antibiotics…I gave them marshmallows and M&Ms [representing bacteria with

different traits]. The toothpick was supposed to simulate the hand sanitizer because it punctures the cell membranes. I gave them 7 seconds and 3 trials. They had to pick up as much bacteria as they could [with the toothpicks] and then, after 3 generations, they would see that only left the M&Ms because…they have that hard shell. And after regeneration, the M&Ms had to reproduce, so then when they graphed that data they can see the exponential growth of the M&Ms and the decline of the marshmallows.

Beth's ability to describe how the various elements of the simulation modeled the factors important to natural selection, and her reference to engaging students through a connection to a familiar phenomenon (taking a course of antibiotics), demonstrated her application of PCK. Through the inquiry cycle, the residents were required to articulate the reasons they selected specific instructional strategies and how those strategies would foster students' understanding of concepts from the curriculum, just as Beth did here.

6.5.3.2   Revising Instructional Strategies Based on Assessment

In these samples of resident performance, PCK is essential to the ability to make an inference about student thinking based on a sample of student work, but it is also reinforced and further developed and refined by testing those inferences through deliberate selection and revision of instructional strategies and assessment after those strategies are implemented. This phase of the work corresponds to the stages of the inquiry cycle in which teachers implement instruction, assess students, use those assessment results to reflect on the impact of the instruction, and then revise their instructional strategies to bring students closer to the learning target. These parts of the inquiry cycle rely on and refine PCK in the areas of planning and implementing instructional strategies that are targeted toward improving student learning of specific content and practices.

All three of the case study residents described at least one example of revising their instructional strategies based on assessment outcomes in order to support student learning, and two of the three made explicit references to how they planned for the future based on student thinking. For example, in her Defense of Learning, Ilana described how she planned a lab activity with students and specifically structured the format for the lab report in order to respond to the challenges her students had demonstrated with using evidence from the data to connect to a scientific concept. She hypothesized that her middle school students might have an easier time using quantitative data to support the idea of inertia around a curve: "They took a marble and went around a track. Then they observed how it left an opening…This is the rubric that I used [for the lab report]…emphasizing more of what's expected within the reasoning section and how should that look. And then, changing to use tests that result in quantitative data so things that were really numbers-based might be a little more concrete for them." In this example, Ilana combined her knowledge of the content and disciplinary practices with her analysis of what had previously been challenging for her students in order to plan an instructional activity and assessment for them that would scaffold their growth on both the content and scientific practice.

In Steven's Looking at Student Work session, a conversation with the program coach elicited his decision around a revision to his instructional strategies based on his diagnosis of the patterns in student difficulty related to the content. In this case, using his PCK related to learning progressions (the fundamental skills and knowledge that build toward understanding of a more complex concept), he was able to identify what students needed to learn and articulate a strategy for supporting them:

Steven: The majority of students had a strong understanding what a redox reaction is. I think it was...where they couldn't identify the oxidation number to begin with, and I think those were the ones who couldn't get [the right answer].
Coach: So that's more like a technical or procedural skill?
Steven: Yeah.
Coach: Okay, so how are you going to help them learn that?
Steven: Definitely provide them a more step by step visualization, and the go-to rules because it's really a dry kind of process to identify oxidation numbers.

The two case study residents who were generally more successful in their inquiry processes each demonstrated at least one example of their ability to describe the effect they believe their instructional strategies had on student learning, although in several cases the connection was at a more superficial level in terms of student understanding of the expectations or standards for effective work rather than conceptual understanding of science concepts. For example, in her Defense of Learning, Ilana noted that both the way her rubric was structured and how she introduced students to writing scientific conclusions were reflected in the challenges students had with their writing:

Seventy percent of students were able to provide the scientific concept that was related but weren't able to connect it with their claims and evidence. So a lot of the times they would just say, oh it's inertia, and then define what inertia was. So their reasoning may have been then, I have provided the concept, which is what was asked of them in the rubric and in what we've gone over. But they weren't making those connections clear for the rest of what they were observing and what that tie to the concepts was.

Here, Ilana takes responsibility for the impact that her instructional strategies had on students' demonstration of their learning, which leads her to plan more explicit instruction related to supporting students in connecting their evidence and observations to scientific concepts. Through this phase of the inquiry process, she refines her PCK around how to plan to support students in developing their scientific writing.

In contrast, Beth, our struggling resident, was not as able to utilize the inquiry process to reflect critically on the impact of her instructional strategies in order to revise them. She analyzed her students' work and came to reasonable inferences about their understandings, but did not progress to the next step of considering how her own choice of teaching "moves" might have influenced the way students learned the material; when asked to describe her next steps, she identified that she was going to "review," "explain," or "teach" the topics students did not yet understand:

What I had to review was that basically all the species that don't reach the top are extinct, because I only had 70 percent of the class that understood that, and explain the relationships between each node on the tree. So [I will] use the term common ancestor more often and basically explain that when something evolves into something else, that species is also considered extinct. My new thing that I'm going to teach is the 4 factors that drive evolution: overproduction, competition, genetic variations, and natural selection.

Beth does use information about students' learning to decide *what* to teach next but not to decide *how* to teach. In other words, she is not refining and developing her personal PCK related to strategies for teaching this particular content – her initial PCK (or lack thereof) remains intact. When compared to Ilana and Steven's students, Beth's students made less progress during the period of her inquiry cycles. This may in part be attributed to her emphasis on reteaching or reviewing topics that were challenging for students, instead of using the inquiry process to refine her PCK and revise her instructional strategies.

6.5.3.3   Analysis of Individual Work Versus Patterns and Trends for Groups

Another major distinction between Beth on the one hand and Steven and Ilana on the other was also apparent in the data. All three case study residents demonstrated different ratios of codes, reflecting that they displayed different frequencies of types of analysis and reflection. However, Beth, who generally struggled the most with the inquiry process and residency year assessments of teaching, engaged in the most analysis of individual student work and the fewest examples of identifying trends and patterns. Ilana and Steven, who performed better in their classroom performance assessments during the residency year, primarily discussed trends and patterns for groups of students and demonstrated less focus on individual students in their discussions of student work, both in the Defense of Learning and in the Looking at Student Work sessions. In order to be successful in making connections between student thinking and instructional strategies and ultimately in revising instructional strategies to be more effective, residents need to generalize and identify patterns and trends, not just focus on individual analysis. It is not practical or doable for novice teachers to differentiate or revise every lesson plan based on the individual needs of 30 students, but it is feasible to align their instruction with the needs of 2–3 groups or respond to 2–3 key patterns. This skill set also relates to the ability to make connections between canonical PCK and personal PCK; when residents can link a generalization from the PCK literature (such as a common misconception) to a trend they have identified in a set of student work, they can then identify a set of actionable responses.

6.5.4   What Types of Structures or Prompts (e.g., Questions, Protocols) Develop Novice Teacher Capacity in Inquiry and PCK?

How do residents successfully develop this complex skill set? In our analysis, we noted content-specific prompts from coaches and peers related to all six of our focus areas of the inquiry cycle. The most frequently appearing type of prompt from a coach or peer was a PA3-coded prompt, related to analyzing student work to evaluate and monitor student understanding of content, concepts, and practices. PA3 prompts led residents to identify trends and patterns in student thinking. PA1 coded prompts (related to using formative assessment) were the next most common category. In response to PA1 prompts, residents described their decisions about planning and implementing appropriate assessments to measure student progress and level of effectiveness of strategies. The prompts from coaches and peers during the Looking at Student Work sessions clearly functioned as scaffolds to support their PCK growth. These prompts were sometimes question-and-answer sequences, but often they reflected collaborative meaning-making where peers or coaches tested their theories and ideas about their inferences and proposed next steps with the case study residents. In these collaborative moments, residents benefitted from a font of PCK that was more extensive than their individual knowledge; these discussions, as well as conversations with their mentors about their inquiry, allowed residents to access and broaden their own PCK through others' contributions.

Examples of more direct coaching through these prompts included:

- Summarizing or restating the resident's thoughts to help clarify a big idea: "it sounds to me like the big goal...is using evidence to explore questions about evolution and change over time..." (coach to Ilana during Looking at Student Work session).
- Prompting continued observations: "And then what else did you notice?" (coach to Ilana).
- Asking a pointed question to draw the resident's attention to implications of trends in the data, "The kids who responded like that [pointing to example], what do you think is going to be their struggle with understanding evolution?" (coach to Ilana).
- Asking a pointed question to focus the resident on student thinking, "Can we go deeper on what do they think about the mechanism for developing resistance? What are they understanding about the mechanism of developing resistance and what are they not understanding about it?" (coach to Beth).
- Clarifying criteria for success: "Can you explain again what you were looking for as a strong answer?" (coach to Steven).

Examples of collaborative meaning-making included:

- Beth's peer sharing an inference about the understanding demonstrated in some student work: "[The student's answer] says if they're reproducing, they will mutate. It makes me think that the student is confusing sexual reproduction and the variation that's produced through crossing over and mixing of gametes with mutation." (peer resident to Beth).
- Reflecting on the selection of effective scenarios or examples to help students understand core science concepts: "[Student work error] to me, was a complication with understanding about statistics and probability...Conceptually, they may not understand what causes disease, or what causes this disease. So this disease might be an infection or might be a typhoon, and then the mathematical probability, statistics issue..." (coach to Ilana).

- Suggesting instructional next steps based on the student data – in this case, students were using scientific terms, but it was unclear whether or not they actually understood what they meant:

Resident 2: You know, maybe that's the next step. Here is a strong answer, what do these terms mean in the context of this answer?

BETH: So are you saying, like variations?

RESIDENT 1: Don't use those words. Tell me what's going on here without using those words. Put it in your own words, and tell me what's happening here.

BETH: So, are you telling me to have them define variation in a context outside of science?

Coach: No. I would say, explain what's happening within this scientific context without using the word variation.

RESIDENT 1: Or using mutation. Just tell me what's happening to these insects. (peers and coach to Beth)

Throughout all steps of the inquiry process, and in all three of the case study residents, we saw evidence of both the use and development of PCK. The inquiry process supported residents' application of PCK by providing a structure for analyzing evidence to test the effectiveness of instructional decisions they made. Our data show that residents used or developed PCK objective A (using formative assessment to continually evaluate and monitor student understanding of content, concepts, and practices) approximately 2.5 times more frequently than PCK objective B (making instructional decisions that are based on student thinking and addressing misconceptions or disciplinary skill gaps). This was more evident in their DOL presentations but also emerged as a pattern in the student work analysis discussions. The residents' comments demonstrated that they utilized canonical PCK to analyze assessments, develop goals for student learning, plan for instruction, and then interpret the results of further assessment to understand the impact of their instructional strategies and how they might revise their strategies for future lessons to respond to student thinking. In addition, implementing the inquiry process allowed residents to build their personal PCK through their reflection on and analysis of their students' learning and their own teaching. Throughout the process, coaches and peers provided prompts that engaged directly with the content the resident was teaching, pushed the resident to deepen his/her thinking, and collaboratively built meaning through integrating PCK from multiple individuals.

6.6 Areas for Further Study

The artifacts of novice teacher learning examined in this case study have provided strong evidence of the links between inquiry and PCK development and demonstrated that using teacher inquiry as a vehicle for building and refining PCK is a promising practice. This examination has also raised several interesting questions for further study that, if investigated, would provide evidence to deepen our understanding of how the power of inquiry protocols and habits of mind could be successfully used in teacher preparation programs to foster PCK development.

As we have seen in the literature and through this case study, PCK and inquiry are complex, multifaceted, and made up of sets of interconnected knowledge and competencies. This study did not address the question of prioritization – whether any of the facets of PCK or the stages of inquiry are more essential than others for novice teacher development. In addition, we described some of the types of prompts or structures that facilitated teacher inquiry, but did not apply a PCK lens to note whether there might be content- or discipline-specific protocols or key questions for inquiry that could further deepen the links between inquiry and PCK. A study that assessed specific elements of PCK, disaggregated the stages of the inquiry cycle, and analyzed them against observable changes in novice teacher practice could inform our understanding of the most effective ways to leverage inquiry for PCK development.

The interactions between canonical and personal PCK throughout the inquiry process raise many questions about how teachers make sense of received knowledge about teaching and their own experiences. When should teachers, especially novice teachers, use their lived experience to refine or revise the canonical PCK they have acquired through graduate coursework or research? How do we build skills to help teachers identify what contextual factors influence the ways in which their personal PCK develops and interacts with canonical PCK? How might we empower teachers as researchers to contribute their personal PCK to the field?

In addition, the notion of collaborative PCK development – using peer and coach or mentor feedback, input, reflection, and prompts to build a shared font of PCK – deserves further exploration to learn how to best help teachers access, share, and collaboratively build those reservoirs of PCK.

Residency programs also provide professional development and ongoing growth opportunities for mentor teachers, and mentors are engaged in supporting their residents through their inquiry process and the development of their PCK. This study did not address any impacts on mentor teacher practice that the inquiry process might catalyze but that growth could be a persuasive part of the value proposition around using an inquiry-based approach in teacher education if it also supported mentors in building and refining their PCK.

Based on resident performance on the Defense of Learning assessment, it is clear that residents are competent beginning action researchers by the end of their residency year. However, it was beyond the scope of this case study to identify whether or not this has a lasting impact on residency graduate practice. Do residents who are prepared using an inquiry-based approach continue to build their PCK through inquiry after they have graduated from their preparation program? If so, in what settings and through what systems?

Finally, teacher preparation should focus on student learning outcomes (defined broadly) as the ultimate measure of the effectiveness of any program. A more in-depth study of the links between the use of inquiry to foster PCK development and outcomes, both in terms of observable teacher practices and student learning outcomes, would help the field to evaluate the impact of these strategies.

References

Ball, D. L., Thames, M. H., & Phelps, G. (2008). Content knowledge for teaching: What makes it special? *Journal of Teacher Education, 59*(5), 389.

Cochran-Smith, M., & Lytle, S. L. (1993). *Inside/outside: Teacher research and knowledge.* New York: Teachers College Press.

Dawson, K. (2014). Teacher inquiry: A vehicle to merge prospective teachers' experience and reflection during curriculum-based, technology-enhanced field experiences. *Journal of Research on Technology in Education, 38*, 265–292.

Duckworth, E. (1986). Teaching as research. *Harvard Educational Review, 56*, 481–495.

Elliott, J. (1988). Teachers as researchers: Implications for supervision and teacher education, Address to the American Educational Research Association, New Orleans, April 1988.

Gess-Newsome, J., & Lederman, N. G. (Eds.). (1999). Chapter 2, The complex nature and sources of teachers' pedagogical knowledge. In *Examining pedagogical content knowledge: The construct and its implications for science education* (pp. 21–22). Dordrecht/Boston: Kluwer Academic.

Grace, M., Rietdijk, W., Garrett, C., Griffiths, J. (2015) Improving physics teaching through action research: the impact of a nationwide professional development programme. *Teacher Development, 19*(4):496–519.

Gudmundsdottir, S. (1987). *Pedagogical content knowledge: teachers' ways of knowing.* Paper presented at the Annual Meeting of the American Educational Research Association. Washington, D.C. (ERIC Document Reproduction Service NO. ED 290 701).

Huffman, D., & Kalnin, J. (2003). Collaborative inquiry to make data-based decisions in schools. *Teaching & Teacher Education, 19*(6), 569–580.

Langer, G., Colton, A., and Goff, L. (2000). Collaborative Analysis of Student Work: Improving Teaching and Learning. Association for Supervision and Curriculum Development, Alexandria, Virginia.

Langer, G. M., Colton, A. B., & Goff, L. (2003). *Collaborative analysis of student work: Improving teaching and learning.* ASCD: Alexandria.

Magnusson, S., Krajcik, J., & Borko, H. (1999). Nature, sources and development of pedagogical content knowledge for science teaching. In J. Gess-Newsome & N. G. Lederman (Eds.), *Examining pedagogical content knowledge: The construct and its implications for science education* (pp. 95–132). Dordrecht: Kluwer Academic.

Morine-Dershimer, G. and Kent, T. (1999). 'The Complex Nature and Sources of Teachers' Pedagogical Knowledge' in Examining Pedagogical Content Knowledge, Dordrecht: Kluwer Academic Publishers.

Park, S., & Oliver, J. S. (2008). Revisiting the conceptualisation of Pedagogical Content Knowledge (PCK): PCK as a conceptual tool to understand teachers as professionals. *Research in Science Education, 38*, 261–284.

Park, S., Jang, J., Chen, Y., & Jung, J. (2011). Is Pedagogical Content Knowledge (PCK) necessary for reformed science teaching?: Evidence from an empirical study. *Research in Science Education, 41*, 245–260.

Robinson, M. A. (2010). School perspectives on collaborative inquiry: Lessons learned from New York City, 2009-2010. Consortium for Policy Research in Education. Retrieved from http://www.cpre.org/images/stories/cpre_pdfs/ci-llreport2010final(nov).pdf.

Rockman et al. (2014). Measuring effective teaching: New Visions for Public Schools–Hunter College Urban Teacher Residency Project, Year 4. (Unpublished report).

Sagor, R. (2009). Collaborative action research and school improvement: We can't have one without the other. *Journal of Curriculum and Instruction, 3*(1), 7–14.

Saunders, W. M., Goldenberg, C. N., & Gallimore, R. (2009). Increasing achievement by focusing grade-level teams on improving classroom learning: A prospective, quasi-experimental study of Title 1 schools. *American Educational Research Journal, 46*(4), 1006–1033.

Shulman, L. S. (1986). "Those who understand: Knowledge growth in teaching." Educational Researcher Feb. 1986: 4–14. (AERA Presidential Address).

Silva, D., and Dana, N. (2001). Collaborative Supervision in the Professional Development School. *Journal of Curriculum and Supervision, 16*(4), 305–321.

Smith, P. S., Esch, R. K., Hayes, M. L., & Plumley, C. L. (2016). *Developing and testing a method for collecting and synthesizing pedagogical content knowledge*. Presented at the 2016 NARST Annual International Conference, Baltimore.

Tabachnick, B. R., & Zeichner, K. M. (1999). Idea and action: Action research and the development of conceptual change teaching of science. Wisconsin Ctr. for Educ. Research, University of Wisconsin-Madison, Madison, WI. *Science Education, 83*(3), 309–322.

Talbert, J. E., Cor, M. K., Chen, P., Kless, L. M., & McLaughlin, M. (2012) Inquiry-based school reform: Lessons from SAM in NYC. Center for Research on the Context of Teaching at Stanford University [Program Evaluation]. Retrieved from http://www.academia.edu/29864629/Inquiry-based_School_Reform_Lessons_from_SAM_in_NYC

Thompson, P. W. (2016). Researching mathematical meanings for teaching. In L. English & D. Kirshner (Eds.), *Handbook of international research in mathematics education* (pp. 435–461). London: Taylor and Francis.

Zargarpour, N. (2005). A collective inquiry response to high-stakes accountability. California Educational Research Association. Retrieved from http://cera-web.org/wp-content/uploads/2009/05/Collective-Inquiry_NZ_CERA-Disting-Paper_2005.pdf.

Chapter 7
Biology Teacher Preparation and PCK: Perspectives from the Discipline

Christine R. Cain and Sherryl Browne Graves

Abstract This chapter examines the integration of pedagogical content knowledge in graduate-level biology courses in a secondary teacher preparation program. Given the residency format of the program, expertise from college faculty in biological sciences and education, science educators from a science museum, and educators working with a high school network combined to review, redesign, implement, and assess PCK infused biology courses. Using the lens of action research, we examine the effectiveness of the course modifications, the context of the residents' clinical experience on this process, and the outcomes for the content course and work with high school students. Lessons learned and suggestions for the future are offered.

Keywords PCK · Secondary teacher preparation · Biology teaching · Learner understanding · Learning progressions · Prior knowledge

7.1 Introduction

The *Mathematics and Science Teacher Education Residency (MASTER)* built upon a successful partnership between Hunter College and New Visions for Public Schools, a support network of secondary schools that created the Urban Teacher Residency (UTR), which has at the center a year of clinical practice with the support of a mentor teacher. The MASTER residency expanded the partnership in multiple ways. The focus of teacher preparation is centered on mathematics and science teacher preparation. The New York Hall of Science was added to the consortium to provide opportunities for residents to expand their pedagogical repertoire in an informal teaching and learning environment. The Hunter College portion of the partnership increased to include, in addition to the School of Education (SOE), the School of Arts and Sciences (SAS) with a co-PI from the STEM disciplines. Overall, the MASTER residency proposed using pedagogical content knowledge (PCK) as a

C. R. Cain (✉) · S. B. Graves
Hunter College, City University of New York, New York City, NY, USA

© Springer International Publishing AG, part of Springer Nature 2018   133
S. M. Uzzo et al. (eds.), *Pedagogical Content Knowledge in STEM*,
Advances in STEM Education, https://doi.org/10.1007/978-3-319-97475-0_7

strategy to knit together the expertise, talents, and resources of the partners to improve teacher effectiveness, retention, and student outcomes. PCK was added to the preparation of the residents and to the professional development of the mentor teachers who support the residents. This chapter focuses on the inclusion of PCK in master's-level biology courses that are offered in the School of Arts and Sciences. These courses are required for the preparation and certification of graduate pre-service biology teachers at Hunter College.

7.2   PCK

The concept of PCK, first articulated by Shulman (1986), includes "an understanding of what makes the learning of specific concepts easy or difficult: the conceptions and preconceptions that students…bring with them to the learning." Moreover, Shulman states that PCK included, "the most powerful analogies, illustrations, examples, explanations, and demonstrations—in a word the ways of representing and formulating the subject so that it is comprehensive for others" (Shulman 1986). Shulman asserted that PCK distinguished the content expert from the expert pedagogue. That is, teachers had additional knowledge that allowed them to know how to share their content knowledge with learners. It is in the interaction among, or an amalgamation of, content knowledge, knowledge of teaching, and knowledge of students or the context that PCK emerges as a characteristic of effective interactions among teachers and students (Gess-Newsome and Lederman 1999). "PCK is a heuristic for teacher knowledge that can be helpful in untangling the complexities of what teachers know about teaching and how it changes over broad spans of time" (Schneider and Plasman 2011).

To explore PCK, different research approaches have been employed. One avenue has been the exploration of the quality and quantity of teachers' content knowledge relative to the presence of PCK in teaching a specific curricular topic or concept. Others have compared the pedagogical practices of novice versus master teachers hoping to determine if greater experience and increased expertise in teaching make it more likely for teachers to use PCK (Krepf et al. 2017). Another strategy has been to focus on the use of particular pedagogical approaches like inquiry (Park and Oliver 2008) or learning progressions (Krepf et al. 2017) as pathways to the development of PCK. In science it is hypothesized that PCK includes knowledge of (1) students' thinking about science, (2) science curriculum, (3) specific pedagogical techniques for teaching science, (4) assessment of students learning in science, and (5) orientation toward teaching science (Magnusson et al. 1999; Park and Oliver 2007; Park and Chen 2012). It has also been suggested that in addition to courses in content and pedagogy, novice teachers need opportunities to apply PCK to the lesson planning process for specific topics (Daehler and Shinohara 2001; Loughran et al. 2012) or to analyze the pedagogical practices presented in videotaped lessons of specific concepts (Roth et al. 2011; Johnson and Cotterman 2015).

7.3   Context of Biology Teaching

Traditionally in our context, the preparation of pre-service biology teachers includes graduate courses in education and in the discipline, in this case, biology. Graduate courses in education are the responsibility of the School of Education, while graduate courses in the discipline, biology, are the responsibility of the School of Arts and Sciences and Department of Biological Sciences. The preparation for secondary school biology teachers relied on courses designed for graduate students seeking a master's degree in biology in molecular and cell biology, cancer biology, molecular and developmental genetics, molecular neuroscience, and biotechnology. Many of the students in the biology master's program are considering doctoral study in the discipline. With the introduction of the MASTER project, several new courses, BIOL 604 PCK I, Special Topics in Biology, and BIOL 605 PCK II, Special Topics in Biology, were added to the disciplinary component of the master's program for biology teachers.

As a part of the MASTER residency, graduate students take graduate-level courses 1 day in a week, and in the remaining 4 days, they are in New York City public high schools. Each resident is matched with a mentor teacher, assigned teaching responsibility for a single period per day during the mentor's five periods per day teaching assignment, and is expected to co-teach and observe the remaining four periods. While it is the intent of the program for the resident to act as the "teacher of record" for that section, the mentor teacher is still responsible for student outcomes in the district's accountability system. In the case of biology, the Living Environment course is mandatory for high school students who are required to take and pass a state-designed (Regents) exam as part of their high school portfolio for graduation.

In designing the new PCK biology courses, the groups worked together. Hunter College biology faculty and education faculty collaborated with science educators from the New York Hall of Science. Members of this group designed, implemented, and examined experimental versions of the courses. College faculty prepared curriculum proposals and shepherded the courses through the complex and lengthy curriculum development process.

7.3.1   Course Goals

PCK: Special Topics in Biology was developed to bolster conceptual understanding of the big ideas in biology for MASTER residents and offer effective strategies to facilitate student learning in science. With a firm grasp on biological content, we wanted residents to develop effective tools to probe students' understanding of biological concepts as well as to identify topic misconceptions. To assess the development of student thinking, residents would learn to create learning progressions to strategically tailor teaching to address their student's current level of understanding.

Given this combination of deep content knowledge and strategies to teach it, we hoped to arm residents with the tools to increase student content knowledge to a grade-appropriate level.

7.3.2 Acquiring a Deep Conceptual Understanding of the Big Ideas in the Discipline of Biology

When teachers hold flawed conceptions of biological phenomena, student learning is negatively affected (Boo and Ang 2004). Topic misconceptions held by the teacher can lead to flawed questions on assessments, which can result in test invalidation. Therefore, we thought it important that residents discover and correct their own misconceptions about biological content. Our residents come from multidisciplinary academic backgrounds. While some hold biology or other scientific degrees, others have taken only a few science courses while majoring in nonscientific disciplines. As a result, we aimed to normalize our cohort's conceptual understanding of content to that of the first-year biology major at the college level and chose content readings from a text used in Principles of Biology at Hunter College.

At this level of study, residents should possess deep conceptual knowledge of topics covered in New York State's core life science requirement, the Living Environment. In this course, high school students use scientific inquiry and mathematical analysis to learn (1) how living and nonliving things differ, (2) how organisms inherit genetic information, (3) how evolution results in organismal change over time, (4) how the continuity of life is sustained through reproduction and development, (5) how organisms maintain a dynamic equilibrium that sustains life, (6) how plants and animals depend on each other and on their physical environment, and (7) how human decisions and activities have impacted the physical and living environment. For high school students, this class sets the stage for further studies in the biological sciences, and strong Living Environment teachers have a positive impact on future success.

7.3.3 Learning to Identify Students' Misconceptions of Biology Topics to Aid the Restructuring of Student Thinking into Scientifically Acceptable Ideas

Early work by Piaget (1929, 1930) laid the foundation for the theory that children understand biological phenomena from their experiences in the world which may later place obstacles to the acquisition of a canonical understanding of scientific content. Other studies have also documented the difficulties students encounter as they learn scientific concepts. For example, in the field of cell biology, students may have difficulty differentiating terms such as cell wall and cell membrane (Díaz de

Bustamante and Jiménez Aleixandre 1998), understanding levels of organization in multicellular organisms (Dreyfus and Jungwirth 1988), or understanding cell processes such as mitosis and replication (Lewis and Wood-Robinson 2000).

To address this challenge, each week our residents were assigned primary literature documenting student misconceptions in science. They were then asked to discuss their own misconceptions uncovered during the readings and why these misconceptions might be reasonable. These discussions, as part of a reflective process, were critical to the exposure of fallacies in our residents' thinking. This experiential learning tended to increase their appreciation for the learning process and the ideas their own students bring to class.

7.3.4 Probing Learners' Understanding of Biological Concepts

The practice of generating effective probing questions was important to our PCK approach. Effective probes allow for the evaluation of students' prior knowledge and provide data for restructuring student thinking into scientifically acceptable ideas. Tofade and colleagues state that "Well-crafted questions lead to new insights, generate discussion, and promote the comprehensive exploration of subject matter. Poorly constructed questions can stifle learning by creating confusion, intimidating students, and limiting creative thinking" (Tofade et al. 2013).

Throughout *PCK: Special Topics in Biology*, residents were tasked with creating probing questions specific to the topic at hand and collecting elicited student data from the field. Each week, a particular modality for data collection was undertaken (e.g., written student work, clinical interviews, small group discussions, and whole group discussions). In class, these data were presented for discussion and analyzed for effectiveness (e.g., if the probe elicited specific knowledge about the topic). If the elicited data revealed a large number of off-target responses, it was scored as not specific enough, edited, reissued, and analyzed again. Fine-tuning probing questions to elicit specific knowledge is a practice even experienced teachers will perform repeatedly. If done well, the practice lends itself to the discovery of topic misconceptions that cannot be overvalued.

7.3.5 Developing Learning Progressions as a Pedagogic Tool to Assess the Growth and Development of Students' Thinking About Topics

The ability to develop learning progressions to rank learners' conceptual knowledge is a useful skill for teachers. Learning progressions serve as a ladder of comprehension about a topic, purposefully constructed to progress from less to more sophisticated understanding or skills. At the top is the highest level of scientific understanding, while the lowest level is informed by prior knowledge or experience. As a tool, learning progressions have shown promise in improving instruction and assessment (Corcoran et al. 2009).

Armed with deep conceptual knowledge, an awareness of biology topic misconceptions, and an understanding of their students' misconceptions elicited through effective questions, we wanted residents to use learning progressions as a scaffold onto which students' prior knowledge can be sorted and used to build up to a canonical scientific understanding. Proficiency at generating learning progressions is challenging, and we found it useful to follow a protocol for their development. The first step begins with a decision about what constitutes a high-level scientific understanding of a particular topic. For example, the topic of cell division would be understood by the college undergraduate at the following level of detail:

Organisms grow by creating more cells through exponential cell division. For cells to divide, they must reach a certain size and replicate their complete set of chromosomes faithfully for distribution into two cells. The timeline of replication and division, known as the cell cycle, is different for every cell type, with neurons dividing most infrequently and epithelial (skin) cells dividing most rapidly. The cell cycle is divided into four stages: DNA synthesis (S) and mitosis (M) with gap (G) phases between them in the following order, G1-S-G2-M. During interphase (G1-S-G2), the cytoplasm and chromosomes are replicated, and chromosomes appear as long, diffuse strands.

Mitosis is also divided into stages. Cells enter prophase with duplicated chromosomes which condense as the nuclear envelope breaks down. Microtubule spindles form which attach to each chromosome, pulling them toward the cell's center. In metaphase, duplicated chromosome pairs align on the equatorial plate and then separate to opposite poles of the cell in the next stage, anaphase. Telophase is characterized by chromosome de-condensation and the reappearance of the nuclear envelope. Lastly, cytokinesis divides the cytoplasm and two daughter cells result.

The cell cycle is tightly regulated by a system of factors known as cyclins and cyclin-dependent kinases (CDKs). Cyclins are so named due to their cyclic expression levels, with specific cyclin levels rising and falling throughout the cell cycle. When a specific cyclin's levels are high, it forms a complex with a more evenly expressed cyclin-dependent kinase. This activates the kinase which allows it to phosphorylate its target and move the cell cycle forward. When the specific cyclin's levels are downregulated, its associated CDK activity ceases allowing for the next cyclin/CDK complex to function.

At critical times in the cell cycle, there are checkpoints where the integrity of the cell and its contents are assessed. At the G1/M checkpoint just before cells replicate their chromosomes, DNA integrity is assessed. At the G2/M checkpoint, just prior to cell division, completion of DNA replication is assessed. A cell must meet the requirements of its checkpoint to move forward in its cycle. If signals generated by double-stranded DNA breaks, or stalled replication forks are present, the cell cycle pauses, and the DNA repair machinery is activated to correct the damage before damaged DNA is propagated to future cells. DNA damage increases the likelihood of uncontrolled cell growth (cancer); importantly, a high percentage of DNA found mutated in tumors are genes expressing members of the cell cycle machinery.

In the protocol's second step, this high-level scientific understanding is separated into subdivisions or threads. For example, the topic of cell division can be divided into the following threads: (1) growth, (2) the cell cycle, (3) mitosis, (4) cell cycle regulation, and (5) repair. Each thread is considered separately when informing individual levels of understanding on the learning progression.

The goal of learning progressions is to assist teachers in meeting all learners at their respective levels of understanding and grow their scientific understanding from there. Therefore, the number of levels chosen for the learning progression and the span of content depend on the topic, the ages of students being taught, and course goals. An example of a learning progression on cell division intended for students with high understanding is shown in Table 7.1. For simplicity, this example has three levels of understanding: high, intermediate, and low. However, given the need to differentiate further, it may be necessary to include more than one intermediate level.

7.3.6 Learning to Use Students' Prior Knowledge to Build New Biological Understanding

Ultimately, we wanted residents to learn to use students' prior knowledge to build new biological understanding. Prior conceptions are known to be resistant to change and can persist in spite of new knowledge delivered in the classroom. This belief perseverance (Savion 2009) is rooted in the theory of cognitive disequilibrium (Piaget 1971), where new information is not taken on because it conflicts with a current worldview. In this state, students tend to misunderstand or discredit new information because it conflicts with what they know to be true.

The combination of generating data with effective questions and distributing the data on learning progressions allows the teacher to determine a student's understanding of the topic. The interpretation of student understanding is based not only upon knowledge of specific scientific concepts but also on the use and accuracy of scientific language, the understanding of relative scale, connections across content areas, and any other criteria deemed relevant by the teacher. Once the data is distributed on the learning progression, the teacher sets goals to increase student knowledge by degree. It may be prudent, for example, to expect students at the lowest level of understanding to reach an intermediate level. Students at an intermediate level might be expected to attain a high level of understanding with instruction. The ability to differentiate lessons for students of all learning abilities is implicit in this pedagogical approach.

7.4 Reflections and Lessons Learned

We used the action research approach to examine the effectiveness of the PCK courses. Through reflective self-study, the faculty associated with the courses examined all aspects of the course to determine what was working and what was not. As Souto-Manning (Souto-Manning 2012) notes, "...action research involves a systematic and sustained study of some aspect of teaching and learning." Using this

Table 7.1 Cell division learning progression

Growth (thread 1)	
High understanding	Students understand that organisms get bigger by creating more cells through the process of cell division. They know that cell growth is exponential and can represent exponential growth with in number series and/or pictograph
Intermediate understanding	Students know that organisms are made of cells but may think that organisms grow by cells getting bigger. If they know that cells divide, there is no understanding of exponential growth
Low understanding	Students understand growth on a macroscopic level as organisms getting bigger because they eat food. If there is knowledge of cells, they think thing that organisms grow because cells get bigger. They may be confused about the difference between cells and cell phones
Cell cycle (thread 2)	
High understanding	Students have detailed understanding of each stage of the cell cycle including the order in which they occur. They know that there are different types of cells with different cell cycles. They can give examples of specific cell types and their relative cell cycle speeds
Intermediate understanding	Students know cells divide; they may think that timing is random or that all cells divide at the same rate and frequency. They may recognize some terminology but cannot explain cell cycle stages, their order, or what is occurring at each stage
Low understanding	Students are unaware that cells exist. If they do know about cells, they may think that all cells are the same and live forever without being replaced
Mitosis (thread 3)	
High understanding	Students understand mitosis as the process of cell division. They know it as a small part of the larger cell cycle which involves DNA and cytoplasmic replication. They know the names and order of stages. They are able to recognize each stage when presented with a picture
Intermediate understanding	Students understand that cells divide but may not be familiar with term "mitosis." they may think that when cells divide, they merely split into two cells. There is no consideration of DNA replication. They may recognize one or two stages when presented with a picture
Low understanding	Students have no knowledge of mitosis or its stages. If they know cells divide, they do not know division as a process
Cell cycle regulation (thread 4)	
High understanding	Students understand the role of checkpoints in maintaining the timing of cellular events. They can name a checkpoint, where it occurs, what is being assessed, and why it is important. They understand cancer as the formation of tumors due to deregulation of cell cycle control
Intermediate understanding	Students have knowledge of cells but are unaware of controls or mechanisms regulating cell division. They may believe that all cells divide randomly. They understand cancer as the uncontrolled growth of cells but do not know why it happens
Low understanding	Students have no knowledge of cells, the cell cycle, or its regulation. They know cancer as tumors and understand them as massive structures that cause people to die. They may associate cancer with hair and weight loss

(continued)

Table 7.1 (continued)

Repair (thread 5)	
High understanding	Students understand that errors can be made during replication and the importance of DNA repair to the integrity of future daughter cells. They can name specific signals that cause repair to take place
Intermediate understanding	Students understand repair on a macroscopic level as the formation of scabs and scars but understand that cells have been injured. They do not understand that errors can occur during cell division that can be fixed. There is little knowledge of DNA replication and therefore no knowledge of specific signals that can signal repair
Low understanding	Students understand repair on the macroscopic level as the formation of scabs and scars. They do not understand that injury and repair happen on the cellular level since they have no concept of cells

approach there were multiple data sources that were used including the course syllabus, individual class sessions and accompanying resources, resident talk and work, and student work collected by residents.

We taught PCK to two cohorts and made significant course design changes between them. We wanted the course to push residents to learn both content and pedagogy at the same time. We assumed that all of our residents possessed at least an undergraduate-level comprehension of biological concepts upon entrance to the MASTER program. Therefore, we programmed a relatively small amount of content instruction (45 min) with most of the time devoted to pedagogy. This differs from other graduate science courses which devote the entire class (150 min) to content instruction.

We quickly discovered that our first cohort required more than 45 min devoted to content presentation. This was not surprising given their varied backgrounds and in some cases, the span of time since their undergraduate years. For example, science majors with chemistry and physics degrees, while knowledgeable about the scientific method and mathematical processes, demonstrated large gaps in their content knowledge across a range of topics in the biology curriculum. Generic scientific knowledge did not bridge or fill those gaps. Residents with nonscientific degrees often found themselves learning particular biological topics for the first time. In the second cohort, a more effective balance was reached with half of the course (75 min) devoted to content presentation, review, or clarification as needed.

From the pedagogical perspective, we observed that even residents with the strongest backgrounds in biology, those that possessed deep knowledge of biological concepts, were challenged by the prospect of teaching content to students with little background knowledge or interest in biology. While these residents had content knowledge expertise, they had not reflected on their own learning or thought about their level of understanding from the perspective of teaching the information to others.

An unexpected challenge we encountered was that the topics residents addressed in PCK, although aligned with the New Visions sequence of course topics, were usually not in sync with topics covered in their mentored classrooms. In addition, the demands of pre-service training and mentor teacher scheduling often left little time for residents to collect student data. Although we affirmed that data could be collected from outside the mentored classroom (e.g., from the kid next door; from students in other classes), some of our residents were never able collect their own data and relied solely on data from fellow classmates. As a result, the data available for analysis and examination in class was not as rich, varied, or representative of different placement settings and student diversity. For these same reasons, residents were not always able to employ their learning progressions in "real-life" teaching. However, we believe our approach to PCK to be an ideal combination of both consolidating scientific concepts for residents and arming them to teach science in a substantive and meaningful manner.

7.4.1 Approaches that Worked

The PCK final project allowed our residents to consolidate each part of the class into one deliverable. They were tasked with the composition of a ten-page paper melding deep content knowledge and the teaching strategies covered in the course. The paper required three major components: (1) a description of the big idea and what counts as a high level of understanding of that key concept, (2) a learning progression for the big idea linked to literature on student thinking about the topic, and (3) an analysis of student thinking resulting from an in-depth probing (one that lasts longer than one exchange) of student thinking about the big idea.

Analysis required residents to use two out of the four modalities (e.g., written work, clinical interviews, small group discussions, or whole group discussions) to gather evidence of student thinking. Instructional context, observations about the target students, inferences, and links to learning progressions were also required. Finally, implications for instruction were elicited by responses to prompts about further assessment and instruction.

This assignment allowed us to determine our success at delivering both science and pedagogical instruction to the class. For residents in our second cohort, the extended content instruction proved beneficial. They demonstrated increased conceptual depth and greater confidence in their scientific knowledge. Residents with scientific degrees had the opposite challenge of encountering difficulty comprehending pedagogical practices but were pushed by the assignment to understand the teaching tools as a whole when applied to a single topic.

7.5 The Future

Overall, we believe our approach to PCK proves useful to any pre-service teacher. Our challenge was an inability to sync our residents' PCK study topics with those in their focus/mentored class.

We postulate two solutions to this problem. Firstly, requests could be made of mentor teachers to identify several topics for which the resident could take responsibility. In this way the resident could develop a few areas of expertise where they could gather student thinking, employ learning progressions, and develop assessment tools for the selected topics. Such an approach would provide residents with more opportunities to attempt PCK approaches with the support of the mentor teacher.

Another second solution might be to create opportunities for low-stakes teaching. One such opportunity might be the development of a summer biology enrichment curriculum offered in collaboration with an informal partner, such as a museum. Residents could design a curriculum recruiting 8th graders from partner middle schools. Mornings could be devoted to residents delivering course content, and afternoons dedicated to reflection with the support of faculty.

In conclusion, it is important for teachers to develop deep content knowledge and pedagogical tools that are relevant to teaching assignment and grade level. Preservice teachers require access to students and opportunities to elicit student thinking about biological concepts. With students' prior knowledge in hand, both novice and experienced teachers are better positioned to build new and accurate biological understanding.

Acknowledgments The authors would like to thank Matty Lau for the overwhelming generosity of her immense knowledge and support during the development of PCK: Special Topics in Biology.

References

Boo, H. K., & Ang, K. C. (2004). *Teachers' misconceptions of science as revealed in science examination papers.* Annual Conference of the Educational Research Association, 2004, Singapore.

Corcoran, T. B., Mosher, F. A., & Rogat, A. (2009). Learning progressions in science: An evidence based approach to reform. CPRE research reports. http://repository.upenn.edu/cpre_researchreports/53.

Daehler, K. R., & Shinohara, M. (2001). A complete circle: Exploring the potential of case materials and methods to develop teachers' content knowledge and pedagogical content knowledge of science. *Research in Science Education, 31,* 267–288.

Díaz de Bustamante, J., & Jiménez Aleixandre, M. P. (1998). Interpretation and drawing of images in biology learning. In H. Bayrhuber & F. Brinkman (Eds.), *What – why – how? research in Didaktik of biology* (pp. 93–102). Kiel: IPN.

Dreyfus, A., & Jungwirth, E. (1988). The cell concept of 10th graders: Curricular expectations and reality. *International Journal of Science Education, 10,* 221–229.

Gess-Newsome, J., & Lederman, N. G. (1999). *Examining pedagogical content knowledge: The construct and its implications for science education.* Boston: Kluwer Academic Publishers.

Johnson, H. J., & Cotterman, M. E. (2015). Developing pre-service teachers' knowledge of science teaching through video clubs. *Journal of Science Teacher Education, 26,* 393–417.

Krepf, M., Ploger, W., Scholl, D., & Seifert, A. (2017). Pedagogical content knowledge of experts and novices—What knowledge do they activate when analyzing science lessons? *Journal of Research in Science Teaching, 55,* 44. https://doi.org/10.1002/tea.21410.

Lewis, J., & Wood-Robinson, C. (2000). Genes, chromosomes, cell division and inheritance – Do students see any relationship? *International Journal of Science Education, 22,* 177–195.

Loughran, J., Berry, A., & Mulhall, P. (2012). *Understanding and developing science teachers' pedagogical content knowledge*. Rotterdam: Sense Publishers.

Magnusson, S., Krajcik, J. S., & Borko, H. (1999). Secondary teachers' knowledge and beliefs about subject matter and their impact on instruction. In J. Gess-Newcombe & N. G. Lederman (Eds.), *Examining pedagogical content knowledge* (pp. 95–132). Dordrecht: Kluwer.

Park, S., & Chen, Y.-C. (2012). Mapping out the integration of the components of pedagogical content knowledge (PCK): Examples from high school biology classrooms. *Journal of Research in Science Teaching, 49*, 922–941.

Park, S., & Oliver, S. (2007). Revisiting the conceptualization of pedagogical content knowledge (PCK): PCK as a conceptual tool to understand teachers as professionals. *Research in Science Education, 38*, 261–284.

Park, S., & Oliver, S. (2008). Revisiting the conceptualization of pedagogical content knowledge (PCK): PCK as a conceptual tool to understand teachers as professionals. *Research in Science Education, 38*, 261–284.

Piaget, J. (1929). *The Child's conception of the world. Harcourt*. New York: Brace.

Piaget, J. (1930). *The Child's conception of physical causality*. London: Kegan Paul.

Piaget, J. (1971). *Biology and knowledge*. Chicago: University of Chicago Press.

Roth, K. L., Garnier, H. E., Chen, C., Lemmens, M., Schwille, K., & Wickler, N. (2011). Video based lessons analysis: Effective science PD for teacher and student learning. *Journal of Research in Science Teaching, 48*, 117–148.

Savion, L. (2009). Clinging to discredited beliefs: The larger cognitive story. *Journal of the Scholarship of Teaching and Learning, 9*, 81–92.

Schneider, R. M., & Plasman, K. (2011). Science teacher learning progressions: A review of science teachers' pedagogical content knowledge development. *Review of Educational Research, 81*, 530–565.

Shulman, L. S. (1986). Those who understand: Knowledge growth in teaching. *Educational Research, 15*, 4–14.

Souto-Manning, M. (2012). Teacher action research in teacher education. *Childhood Education, 88*, 54–56.

Tofade, T., Elsner, J., & Haines, S. T. (2013). Best practice strategies for effective use of questions as aTeaching tool. *American Journal of Pharmaceutical Education, 77*, 155.

Chapter 8
Pedagogical Content Knowledge in a Mathematics Adolescent Education Master of Arts Program: A Case Study

Robert Thompson

Abstract This chapter is a case study examining the symbiotic relationship among an Arts and Sciences mathematics department, a school of education, and a large urban public school district, as a context for the development of pedagogical content knowledge in the curriculum. Specifically, Hunter College, CUNY, has a long-standing master's program in adolescent mathematics education, which is a joint program with the Department of Mathematics and Statistics and the School of Education. Recently a subset of students in this program have also been involved in a teacher education residency program funded by the National Science Foundation. One of the goals of this residency program is to explore various interpretations of pedagogical content knowledge and their implementation in the college and the high school curriculum. This chapter describes several of the courses in this master's program, which seem to be fertile grounds for pedagogical content knowledge. We consider the creation of a new course, dealing with misconceptions in mathematics learning, which explicitly revolves around the idea of PCK, a geometry course which was revised and optimized for the teacher education master's program and incorporates PCK, and an existing problem-solving course, which naturally contains elements of PCK.

Keywords PCK · Adolescent mathematics instruction · Teacher residency program · Pre-service learning · Misconceptions · Axiomatic geometry

8.1 History of the Adolescent Education Master's Program in Mathematics

The Department of Mathematics and Statistics is a department in the School of Arts and Sciences at Hunter College, one of the senior colleges in the City University of New York. Hunter has bachelor's, master's, and a few Ph.D. programs. Mathematics

R. Thompson (✉)
Department of Mathematics and Statistics, Hunter College City University of New York, New York, NY, USA

© Springer International Publishing AG, part of Springer Nature 2018   145
S. M. Uzzo et al. (eds.), *Pedagogical Content Knowledge in STEM*,
Advances in STEM Education, https://doi.org/10.1007/978-3-319-97475-0_8

and statistics has a bachelor's degree in mathematics, bachelor's degree in statistics, and several master's programs, including one in Mathematics Adolescent Education. The students in this program are preparing to teach mathematics at the middle school or high school level, typically in a New York public school.

This degree program is run jointly with the School of Education at Hunter. The students in the program must complete 41 credits of course work, 26 credits of which are in the SOE, and 15 are in A&S. The 15 credits in A&S consist of 5 3-credit courses in our department. These courses are as follows:

- MATH 620 secondary school mathematics from an advanced perspective I.
- MATH 630 secondary school mathematics from an advanced perspective II.
- MATH 633 axiomatic geometry.
- MATH 635 problem-solving explorations OR MATH 655 challenging concepts in mathematics: Using research to identify common misconceptions and assess student learning.
- STAT 612 discrete probability OR STAT 614 data analysis using statistical software.

There are 13 different adolescent education master's programs at Hunter, all joint with A&S and SOE. There are programs in English, Chemistry, Biology, History, Physics, Earth Science, Mathematics, Mathematics Professional Certificate, Social Studies, Chinese, Italian, Latin, and Spanish. The mathematics education master's program is one of the oldest on campus. The current program was established in the 1980s. In the early and mid-part of the 20th century, the mission of the mathematics department at Hunter had primarily been to train prospective teachers. The mathematics department was led for many years by Mary P. Dolciani, a very well-known mathematics educator who wrote over 100 middle school and high school arithmetic and algebra textbooks, which were used all over the country for years. The mathematics education master's program is rooted in this departmental tradition of teacher training.

All five of the required mathematics and statistics content courses are designed specifically for the students in the adolescent education master's program and are taken only by those students. This is in contrast to some of the other adolescent education master's programs at Hunter in which only a few of the 15 content credits involve courses designed specifically for the teacher trainees. Thus this is a significant commitment and involvement on the part of the Mathematics and Statistics Department to the math education master's program.

For many years there have been a number of residency programs associated with the mathematics education master's program. Typically, there has been one such residency program at a time involving mathematics. Some subset of students matriculated in the master's program would also be in the residency program, and the residency program would provide some level of financial support and require a certain amount of in-class teaching experience in addition to the student taking courses and comprehensive exams at the College. These residency programs would also provide some sort of placement mechanism to facilitate the student getting a teaching job upon completion of the program. From 2013 to 2017, the residency program linked to the math education master's program was MASTER – Mathematics and Science Teacher Education Residency.

8.2 Brief Summary of the MASTER Residency Program

The MASTER program was based on a Mathematics and Science Partnership grant from the National Science Foundation and involved a consortium of three institutions – Hunter College CUNY, the New York Hall of Science, and New Visions for Public Schools. It was a 2-year residency program, and there were three cohorts – 2013–2015, 2014–2016, and 2015–2017. The students in the MASTER program selected a discipline – either mathematics, chemistry, earth science, or biology. In addition to satisfying the course work, comprehensive exams, and other requirements of their MA degree program, the students were placed in a participating public school, assigned a teacher-mentor and taught a regular class in the school as the teacher of record.

One of the fundamental goals of the MASTER residency program was to explore and implement pedagogical content knowledge in the curriculum and in the training of the students/prospective teachers. The three courses in the mathematics adolescent education master's program described below were chosen for their potential use of PCK. The first course, on misconceptions in mathematics, was designed in conjunction with the design of the MASTER residency program, and PCK, while not always referred to by that name, is built into the plan of the course. The second course, on problem-solving, is a course that preexisted the MASTER program, was not modified in any way directly related to MASTER, but was taken by MASTER students, and embodies aspects of PCK, again with the PCK not necessarily called out by name. The third course, on geometry, is a pre-existing, in fact a long-standing course in the master's program, which was overhauled shortly after the start of MASTER, with some of the goals of MASTER in mind.

In this chapter we did not explicitly attempt to quantify or assess the PCK in these three courses. The MASTER program does involve significant efforts to assess PCK and that is addressed in other chapters in this volume. Also there is a lot of contemporary research being done on measuring and assessing PCK in the literature, for example, see (Lim and Guerra 2013). Here we are simply documenting the modification of our curriculum in ways that were informed by a desire to enhance PCK.

8.3 MATH 655 Challenging Concepts in Mathematics: Using Research to Identify Common Misconceptions and Assess Student Learning

During the 2012–2013 academic year, as we were designing the MASTER residency program, we approached several faculty members in the Department of Mathematics and Statistics seeking their participation. We were particularly interested in assessing and enhancing PCK in the Mathematics Adolescent Education program. A faculty member who has been working in the field of mathematics education for some time, and has been involved in a number of ongoing programs involving graduate level math education courses for in-service teachers, decided to use this as an opportunity to start a course in misconceptions in mathematics, a topic area that he has been researching for a while (see Cherkas 1992). A new course was created, titled "Challenging Concepts in Mathematics: Using Research to Identify Common Misconceptions and Assess Student Learning," now numbered MATH 655, which the Department offered in the fall of 2013. The students enrolled in the course included the mathematics students in the first cohort of MASTER.

There is a growing body of research in the field of mathematical misconceptions, and the link with PCK is known. For example, in (Chick and Baker 2005) it is shown that looking at teachers' responses to students' misconceptions can be used as a device to measure PCK in a class. More deeply than that, students' misconceptions go to the heart of what makes learning a subject easy or difficult and therefore is central to even the formulation of the definition of PCK. Quoting from Shulman's 1986 seminal paper in which he formulates and presents the concept of PCK, as part of a classification scheme of knowledge relevant to teaching (Shulman 1986), Shulman says "Here, research on teaching and learning coincide most closely. The study of misconceptions and their influence on subsequent learning has been among the most fertile topics for cognitive research. We are gathering an ever-growing body of knowledge about the misconceptions of students and about the instructional conditions necessary to overcome and transform those initial conceptions. Such research-based knowledge, an important component of the pedagogical understanding of subject matter, should be included at the heart of our definition of needed pedagogical knowledge."

Students' progress in studying mathematics is often hampered by layers of misconceptions or misunderstandings regarding a variety of mathematical facts, procedures, and relationships. Traditionally these notions are dismissed by teachers as simply being wrong and are seen as something to disregard. However, mathematics education researchers and practitioners have learned that these misconceptions themselves can be analyzed, tabulated, and classified and often lead to a greater understanding of students' learning processes. This in turn can help the students' progress in ways that can be more effective than just purging mistakes and imparting correct answers. Students' misconceptions are not necessarily the result of irrational thinking and often satisfy a definite internal logic, albeit possibly misapplied to a given situation, and recognition of this can be validating to the student and can help the student appreciate that mathematics is not just some artificial body of random "facts" to be memorized.

MATH 655 is organized around three threads. In the first thread, the students in the class will be become familiar with the literature on misconceptions and the research methodologies employed to study it. The student will formulate a list of common misconceptions, which they can use in their own classroom teaching. In the second thread, they will engage in misconception research by focusing on one particular misconception and devising a lesson plan to approach it. They will draft a research proposal and present a brief paper. The third thread consists of presenting the results of their research. These three components are worth 10%, 15%, and 35% of their grade, respectively. Another 20% of the course grade is based on class participation and another 20% on biweekly written critiques of research articles. The course has one required textbook, *Why Is Math So Hard for Some Children? The Nature and Origins of Mathematical Learning Difficulties and Disabilities*, edited by Berch and Mazzocco (Berch and Mazzocco 2007). There is also an extensive bibliography of required and recommended reading.

8.4 MATH 635 Problem-Solving Explorations

MATH 635 *Problem-Solving Explorations* has been a graduate course in the department for the mathematics adolescent education master's students for quite some time. This course has a hands-on, active learning format with no lecturing. Students take turns working out problems on the board in front of the class, sometimes in collaboration with one another and the class.

The objectives of the course include having the students learn problem-solving techniques and heuristics to solve mathematical problems, including nonroutine and challenging problems. They also learn to identify and implement strategies to help others to solve mathematical problems. They learn how to know when they have a valid solution to a problem, how to justify their results, and communicate both the result and the reasoning to others. In addition, the students become familiar with problem-solving research and identify appropriate problems and problem-solving activities for their classrooms.

This course has been offered for a number of years and predates the MASTER teacher residency program. Like mathematical misconceptions, helping students develop problem-solving skills in mathematics deals with not just mathematical content but understanding what makes certain problems easy to solve, others hard, how to know the difference, and how to think about a mathematical problem. This is fundamental to PCK, and therefore it was natural to include this course in the MASTER program. The students in the second cohort, starting in Fall 2014, took this course as one of their five math content courses, instead of MATH 655, the misconceptions course.

The course grade was based 12% on class participation, 13% on original problems, 40% on solutions to assigned problems, 10% on a research presentation, and

25% on homework. There are no exams. There is one required text for the course, *Problem Solving Strategies: Crossing the River with Dogs*, Johnson, Herr, and Kysh (2012). Here is a list of specific mathematical objectives of the course, as they relate to standards associated with the Common Core.

Standard 1: Knowledge of Mathematical Problem-Solving—Candidates know, understand, and apply the process of mathematical problem-solving.

Indicators:

1.1 Apply and adapt a variety of appropriate strategies to solve problems.
1.2 Solve problems that arise in mathematics and those involving mathematics in other contexts.
1.3 Build new mathematical knowledge through problem-solving.
1.4 Monitor and reflect on the process of mathematical problem-solving.

Standard 3: Knowledge of Mathematical Communication—Candidates communicate their mathematical thinking orally and in writing to peers, faculty, and others.

Indicators:

3.1 Communicate their mathematical thinking coherently and clearly to peers and faculty.
3.2 Use the language of mathematics to express ideas precisely.
3.3 Organize mathematical thinking through communication.
3.4 Analyze and evaluate the mathematics thinking and strategies of others.

Standard 4: Knowledge of Mathematical Connections—Candidates recognize, use, and make connections between and among mathematical ideas and in contexts outside mathematics to build mathematical understanding.

Indicators:

4.1 Recognize and use connections among mathematical ideas.
4.2 Recognize and apply mathematics in contexts outside of mathematics.
4.3 Demonstrate how mathematical ideas interconnect and build on one another to produce a coherent whole.

Standard 5: Knowledge of Mathematical Representation—Candidates use varied representations of mathematical ideas to support and deepen students' mathematical understanding.

Indicators:

5.1 Use representations to model and interpret physical, social, and mathematical phenomena.
5.2 Create and use representations to organize, record, and communicate mathematical ideas.
5.3 Select, apply, and translate among mathematical representations to solve problems.

Standard 8: Knowledge of Mathematics Pedagogy—Candidates possess a deep understanding of how students learn mathematics and of the pedagogical knowledge specific to mathematics teaching and learning.

Indicators:

8.6 Demonstrate knowledge of research results in the teaching and learning of mathematics.

8.8 Demonstrate the ability to lead classes in mathematical problem-solving and in developing in-depth conceptual understanding and to help students develop and test generalizations.

Standard 9: Knowledge of Number and Operation—Candidates demonstrate computational proficiency, including a conceptual understanding of numbers, ways of representing number, relationships among number and number systems, and meanings of operations.

Indicators:

9.2 Use properties involving number and operations, mental computation, and computational estimation.

9.5 Apply the fundamental ideas of number theory.

9.6 Make sense of large and small numbers and use scientific notation.

9.10 Demonstrate knowledge of the historical development of number and number systems including contributions from diverse cultures.

Standard 10: Knowledge of Different Perspectives on Algebra—Candidates emphasize relationships among quantities including functions, ways of representing mathematical relationships, and the analysis of change.

Indicators:

10.1 Analyze patterns, relations, and functions of one and two variables.

10.4 Use mathematical models to represent and understand quantitative relationships.

Standard 11: Knowledge of Geometries—Candidates use spatial visualization and geometric modeling to explore and analyze geometric shapes, structures, and their properties.

Indicators:

11.3 Analyze characteristics and relationships of geometric shapes and structures.

11.4 Build and manipulate representations of two- and three-dimensional objects and visualize objects from different perspectives.

Standard 13: Knowledge of Discrete Mathematics—Candidates apply the fundamental ideas of discrete mathematics in the formulation and solution of problems.

Indicators:

13.1 Demonstrate knowledge of basic elements of discrete mathematics such as graph theory, recurrence relations, finite difference approaches, linear programming, and combinatorics.

13.2 Apply the fundamental ideas of discrete mathematics in the formulation and
       solution of problems arising from real-world situations.

   *Standard 15: Knowledge of Measurement—Candidates apply and use measure-
ment concepts and tools.*

   Indicators:

15.1 Recognize the common representations and uses of measurement and choose
       tools and units for measuring.
15.2 Apply appropriate techniques, tools, and formulas to determine measurements
       and their application in a variety of contexts.

   More than any of the other courses in the master's program, this course strictly
adheres to an active learning model, in which *all* classroom activity involves interaction, discussion, and presentation. That makes this course a fertile environment
for the exploration of PCK. The students aren't just learning how to solve mathematical problems; they are studying and exploring the psychological process of
mathematical problem-solving and learning how to get others engaged in this process, which is a key component of PCK.

8.5   MATH 633 Axiomatic Geometry

This is the oldest of the three courses discussed in this chapter. Several decades ago
this course existed in the department in the form of an introduction to axiomatic
geometry and presumably was taken by mathematics education students and undergraduate mathematics majors. Eventually the course evolved into some version of
differential geometry and algebra and became a pure mathematics course taken
exclusively by upper division undergraduate math majors and pure math masters
students. Roughly 10 years ago, when the department was overhauling the curriculum for the Master of Arts in Mathematics Adolescent Education, it was decided
that it would be good to have an axiomatic geometry course for that program. There
was also a notion that the course might be of interest to pure math majors as well,
both undergraduate and graduate.
   Initially, the course was designed around *Euclidean and Non-Euclidean
Geometries*, by Greenberg (2008). This classic text, which has no equal in the
canon, is a beautiful and deep introduction to axiomatic geometry. It takes as its
central point of view the history of attempts to prove the parallel postulate, one of
Euclid's original axioms, and the one that was long thought to be of a different
character than the others. For centuries mathematicians and logicians attempted to
prove the parallel postulate from the other axioms of Euclid. Greenberg introduces
axiomatic geometry in a rigorous fashion while interleaving it with this fascinating
historical story. Finally, in the 19th century, Lobachevsky and others proved that the
parallel postulate is in fact independent of the other axioms, by exhibiting models in
which all the other axioms hold, but the parallel postulate does not. Therefore, if one

assumes all of Euclid's axioms except the parallel postulate and replaces the parallel postulate with a different assumption, the result is the so-called non-Euclidean Geometry. The two main flavors of non-Euclidean geometry that result from this are called hyperbolic and elliptic geometry.

This is a fascinating story, and no one tells it better than Greenberg. Along the way the students develop solid skills in theorem proving as well as an introduction to hyperbolic geometry, which is important if they go on to graduate study in mathematics.

Unfortunately, this is all rather far afield from a course that would best serve pre-service teachers. There is axiomatic geometry in the Common Core State Standards and the New York State curriculum of course, but not non-Euclidean geometry, and there is not so much emphasis on theorem proving. The audience best suited to the course would have been the pure math bachelor's and master's students, but ironically they were not taking the course at all. The mathematics education master's students were required to take it, and it was a barrier. Furthermore, the instructors typically assigned to the course had no experience with secondary education or teacher training. The course involved little PCK. So in 2013 we embarked on an overhaul of MATH 633. The goals were to:

1. Optimize the course to best serve the needs of pre-service mathematics teachers.
2. Align the content of the course with the geometry portion of the New York state high school math curriculum.
3. Select an appropriate textbook.
4. Introduce PCK into the course.

The department decided to overhaul the course according to the abovementioned goals. The concept of PCK was new to us. Aligning the course with the NYS secondary math curriculum is a challenge because this curriculum, which is ultimately driven by the NYS Regents Exams, is quite the moving target. From 1977 to 2004, the 3-year high school math curriculum in New York corresponded to the Regents Exams, labeled Sequentials I, II, and III. Each exam in this sequence covered a variety of different topics. To some extent the curriculum jumped from one topic to another. From 1999 to 2010, the Regents Exam became two exams, named Math A and Math B, with each one covering 1.5 years of material (note the overlap with the Sequential). In 2008, the state phased in another three-exam sequence, called Integrated Algebra, Geometry, and Algebra 2/Trigonometry. This sequence followed a more traditional arrangement of subject matter that was common in high school curricula prior to the age of the Sequential exams. More recently, beginning in 2014, the state phased in a new sequence Algebra I, Geometry, and Algebra II. This latest sequence still follows the traditional 3-year curriculum but now incorporates the Common Core State Standards in Mathematics. Also, the third exam has far less trigonometry than before and more statistics.

The textbook chosen for MATH 633 is *College Geometry, A Discovery Approach*, second edition, by David C. Kay (2001). This book has been used by a number of other colleges in math BA and math teacher preparation programs. The book is designed to be appropriate for secondary teacher preparation and has secondary school geometry curricula in mind. In addition, there are "moments of discovery" in each chapter which are examples intended to guide the student to discover results on their own. These sections are well suited for the instructor to use help the students gain experience his how to explore geometry with their own high school students.

The course is still fairly proof based, because that is the nature of the high school geometry curriculum. It is a historical fact that high school geometry is based on proofs much more than other subject areas like algebra and calculus, leaving some students with the false impression that proofs are relevant only in geometry. It can be argued that this emphasis on proofs encourages PCK because the instructor isn't just imparting facts (content knowledge) but rather is exploring the reasons for the truth of these facts. However, proofs can be presented in as rote a way as any mathematical topic, so the relationship between mathematical proofs and PCK is not really clear. Nevertheless, geometry classes, and geometry teachers, have been an active area of study for researchers in PCK, and variants of this idea, no less so than other mathematics subjects. See, for example, (Herbst and Kosko 2012).

8.6 Conclusions

The MASTER residency program formally introduced the concept of PCK to a number of us in the department. The concept of PCK seemed a little hard to pin down for myself, colleagues in my department, and my colleagues working on the MASTER residency program. Indeed, those of working on MASTER spent a great deal of time discussing and trying to elucidate what PCK is, and how to identify it, assess it, and encourage it.

It seems to this author that PCK, while not known by or referred to by that name, has existed as a long-standing tenet in the work of colleagues in our department and has played a role in our Adolescent Education Master's program for a long time. MASTER led us to identify it, call it out by name, and thereby support it and enhance it.

The misconception course is based on a subject area that is explicitly part of PCK. The problem-solving course, while not explicitly based on PCK, is a natural place to look for live examples of it. The axiomatic geometry course is an example of an existing course that was remodeled with PCK in mind.

It is interesting to note that during the last decade, there has been some effort on the part of the institution to separate, or draw a line, between the mathematical content knowledge presented to the math education master's students through the five content courses offered by our department, and the pedagogy, both practical and theoretical, that they receive through the School of Education courses. It seems as if there is administrative pressure to separate the two, building a wall between them so to speak, which runs counter to the philosophy that the master's program has followed. The emphasis on PCK in the MASTER residency program has spotlighted this issue. If one believes that PCK is beneficial to teacher education and student learning, then the developments discussed here could be seen as beneficial.

Acknowledgments The author thanks Prof. Barry Cherkas, Prof. Sandra Clarkson, and Dr. Patrick Burke of the Department of Mathematics and Statistics for help in presenting this case study. They are the bedrock of a long-standing and successful mathematics education Master of Arts program at Hunter College. I would also like to thank Sarah Arpin for doing an excellent job of reinventing the Axiomatic Geometry course. I thank my colleagues on the NSF MSP grant which developed the MASTER residency program. This work was supported by National Science Foundation—Mathematics and Science Partnership Grant #1238157.

References

Berch, D. B., & Mazzocco, M. M. (2007). *Why is math so hard for some children? The nature and origins of mathematical learning difficulties and disabilities*. Baltimore: Paul H. Brookes Publishing Co.

Cherkas, B. (1992). The art of undoing wrong mathematics in the classroom. *Primus, 2*, 1–8.

Chick, H. L., & Baker, M. K. (2005). Investigating teachers' responses to student misconceptions. In *Proceedings of the 29th conference of the International Group for the Psychology of mathematics education* (Vol. 2, pp. 249–256). Melbourne: PME.

Greenberg, M. J. (2008). *Euclidean and non-Euclidean geometries, development and history* (4th ed.). Macmillan: Freeman, W.H. & Company, New York, NY.

Herbst, P., & Kosko, K. (2012). Mathematical knowledge for teaching high school geometry. In *Proceedings of the 34th annual north American chapter of the International Group for the Psychology of mathematics education*. Kalamazoo.

Johnson, K., Herr, T., & Kysh, J. (2012). *Problem solving strategies: Crossing the river with dogs* (2nd ed.). Wiley Publishing: Hoboken, NJ.

Kay, D. C. (2001). *College geometry, a discovery approach* (2nd ed.). Addison Wesley Longman, Inc., Chicago, IL.

Lim, W., & Guerra, P. (2013). Using a pedagogical content knowledge assessment to inform a middle grades mathematics teacher preparation program. *Georgia Educational Researcher, 10*(2), 1.

Shulman, L. S. (1986). Those who understand: Knowledge growth in teaching. *Educational Researcher, 15*(2), 4–14.

Chapter 9
Evaluation of PCK in STEM Residency Programs: Challenges and Opportunities

Kay Sloan, Alison Allen, Kristin M. Bass, and Erin Milligan-Mattes

Abstract From 2013 to 2017, Rockman et al. developed and tested various strategies for assessing the pedagogical content knowledge (PCK) of novice and experienced STEM teachers. The research, conducted as part of Rockman's external evaluation of the *Mathematics and Science Teacher Education Residency (*MASTER*)* project, began with a review of existing tools, followed by a pilot study with metrics built around authentic student work. The metrics, customized for math and science teachers, asked teachers to analyze students' learning needs and offer ways to address them. While the research revealed differences within and across disciplines—in, for example, how teachers prioritize procedural and content knowledge—it also confirmed that PCK was a multifaceted, often elusory concept, variously defined by discipline, subject, classroom context, and topic. As researchers explored ways to create reliable tools that captured the nuances of PCK but still generated meaningful data that could inform MASTER development, partners continued to look for ways to fold PCK into teacher training in meaningful ways. Their development of specific objectives for what teachers with strong PCK do in the classroom allowed researchers to examine PCK *situated in practice*, using not a single tool but multiple methods. Analyses of lesson plans over residents' clinical year and retrospective interviews with MASTER-trained teachers, in their induction year, provided complementary opportunities to explore instantiated PCK, defined by both classroom practice and teacher readiness. The lessons learned confirm the challenges of assessing PCK within and between subject areas, as well as some opportunities for research and development around teacher training.

Keywords PCK · STEM · Teacher residency program · Evaluation · Assessing PCK

K. Sloan (✉) · A. Allen · K. M. Bass
Rockman, Bloomington, IN, USA

E. Milligan-Mattes
YES Prep Public Schools, Houston, TX, USA

© Springer International Publishing AG, part of Springer Nature 2018   157
S. M. Uzzo et al. (eds.), *Pedagogical Content Knowledge in STEM*,
Advances in STEM Education, https://doi.org/10.1007/978-3-319-97475-0_9

9.1   Background

From 2013 through 2017, Rockman et al., an independent research and evaluation firm, served as the external evaluator for the *Mathematics and Science Teacher Education Residency* (MASTER) project, a collaborative teacher preparation effort funded by the National Science Foundation's Math and Science Partnership program. MASTER's partners included the New York City Department of Education (NYCDOE); Hunter College; New Visions for Public Schools, a school support organization serving a 70-school network; and the New York Hall of Science (NYSCI). During MASTER's four years, partners refined a 24-month teacher preparation program during which math and science residents apprenticed for a year in the city's high schools under the guidance of a trained mentor, engaged in inquiry-based learning in NYSCI's museum setting, gained skills in New Visions' formative assessment and data-driven approach to instructional practice, and completed Hunter coursework leading to a Master's degree—all with coaching and support that extended into their first, post-residency year of teaching.

MASTER's design called for partners to integrate PCK throughout residents' training and mentors' professional development. In the early stages of the project, partners defined PCK broadly—as a range of knowledge, skills, behaviors, and beliefs that could be incorporated into various elements of training—which allowed each partner to draw on their unique expertise and experience. For example, in rethinking their math and science curriculum and residents' course maps, Hunter College turned to College of Arts and Sciences and School of Education faculty members who had a history of cross-discipline collaboration or practitioner-oriented research interests. New Visions explored ways to build on their inquiry model and emphasis on using data and student work and students' thinking as they designed residents' training and mentors' professional development. Introducing design thinking and inquiry models to residents, in a museum setting with rich opportunities for hands-on inquiry, NYSCI explored ways to help residents develop their students' conceptual understanding.

This collective approach gave the research team multiple paths through which to explore PCK, as it was variously defined and manifested in an array of artifacts. Our initial plans called for fairly standard research procedures—scouring the PCK literature for existing tools with established evidence for validity and reliability, and, as needed, piloting new or adapted tools, developing scales, and testing the new measures with MASTER's three cohorts.

Challenges in relying on existing tools became apparent early on. Our initial reviews of instruments that might serve as a starting point showed that the most commonly used PCK measures, such as the Mathematics Knowledge for Teaching (MKT) and ATLAST assessments, were intended for elementary and middle school teachers (Hill et al. 2004; Smith and Banilower 2006). Searches through various compendia and databases turned up no comparable instruments for high school math and science teachers. Using middle school tests was an option but came with some risks. Using MKT and ATLAST items to measure PCK among apprentices in a high school STEM teacher residency program, UCLA researchers found no significant pre-post differences in PCK—in part because the tests weren't adequately aligned to program content and goals (Wang et al. 2013).

The challenges lay not just in aligning measures to program content and goals but also in aligning them to the larger body of literature on PCK. Even when Shulman first introduced the term, in an address at the 1985 annual meeting of the American Educational Research Association (Shulman 1986), the PCK umbrella encompassed a large, overlapping group of skills—as well it should, as an explication of all teachers need to know and do to be effective. Over the years since Shulman's work, the PCK umbrella coverage has seemed to grow. Magnusson, Krajcik, and Borko (1999), for example, define science PCK as the product of knowledge of subject matter, pedagogy, and context, along with teacher dispositions or orientations toward teaching science, knowledge of curricula, subject- and topic-specific strategies, students' understanding, and assessment. Ball, Thames, and Phelps (2008) add knowledge of conceptions and misconceptions to knowledge of content and of students. At a 2012 summit of PCK researchers convened at the Biological Sciences Curriculum Study, participants generated a consensus definition that, in part, differentiated between tacit and explicit PCK, and between teachers' "personal" PCK and a "canonical" knowledge base that could be integrated into curriculum and professional development (Gess-Newsome 2015). Sean Smith and colleagues at Horizon Research reiterate the concepts of personal and canonical PCK, arguing that the former is experience-supported while the latter is derived from educational research (Smith et al. 2017).

In addition to defining a multifaceted concept, recent research also acknowledges the complexity of measuring PCK, suggesting that metrics be multifaceted as well. Researchers at UCLA, for example, concluded that "teacher knowledge is a complex construct to measure" and that "no one assessment can accurately measure it in its entirety. Instead a battery of well-developed tests is needed to measure pedagogical content knowledge of teachers" (Buschang et al. 2012, p. 24).

Our challenges were twofold: as researchers, we were charged with developing measures or relying on the growing body of research to adapt measures to gauge novice and experienced teachers' PCK. As evaluators, we were also charged with providing MASTER partners with feedback on how well their efforts were helping both groups of teachers develop PCK. In both roles, it gradually became clear that we would need to reconcile the disparity between a big, multifaceted, concept and the need to assess a single facet or manageable set of attributes aligned to MASTER's content and goals, and to the critical components in each partners' emerging framework and training activities.

Our forays into assessing PCK fell into three phases, conducted with MASTER's three cohorts of residents and mentors. In the first phase, we focused on PCK broadly defined, and in an initial pilot and subsequent, larger-scale effort, administered a multipart assessment to gauge how teachers understand and apply PCK. We narrowed our focus in Phase 2, using a multi-method approach to identify common themes across data sources. For the final phase, we examined PCK from a different angle, shifting from an assessment of teacher outcomes to an assessment of training inputs.

9.2 Phase 1: The Big Picture–Assessing PCK, Broadly Defined and Applied

9.2.1 Initial Pilot

At the formal kickoff in early 2013, we began the process of aligning our assessments to MASTER goals, testing a few strategies that might make up our battery of tests. We took our cues from partners' activities and their shared, if variously framed, emphasis on students' work and students' thinking. We created two tasks that asked 17 experienced teachers from the first cohort of project mentors to analyze student work and anticipate student misconceptions—both skills emphasized in MASTER's resident and mentor training. Our samples of student work, provided by New Visions, included the work of actual NYC high school students. We also asked a larger group of math and science teachers to complete a brief survey with items about beliefs and practices, similarly mapped to the project's goals and training, as well as the new Common Core State Standards (CCSS) for math and the Next Generation Science Standards (NGSS).

Among the survey respondents ($N = 50$; 25, math; 25, science), there was general consensus about project goals and PCK-related practices: in response to survey items, math and science mentors alike agreed that instruction should help students grasp the big ideas as well as learn basic facts or methods (math: $M = 5.40$, $SD = 0.89$; science: $M = 5.56$; $SD = 0.70$ on a six-point Likert scale). They reported relatively high levels of confidence and use of project-related strategies, such as "taking students' prior understanding into account when planning instruction" ($M = 3.47$ on a four-point confidence scale; $SD = 0.67$) and "using knowledge of common student misconceptions about content to guide instruction" ($M = 3.47$ on a four-point frequency scale; $SD = 0.61$). Teachers also said that they routinely asked students to "explain their reasoning orally or in writing" ($M = 3.40$; $SD = 0.53$).

Responses from the smaller sample of 17 teachers who analyzed student work also indicated that most could spot the point where students were getting tripped up. What varied were teachers' views on what caused the confusion and what misconception or skill deficit they would address.

- Teachers attributed students' confusion, variously, to misconceptions, computational errors, misreading of graphs, even to the way questions were worded.
- Teachers were also divided on the skills students should learn, and in what order. Some math teachers, for example, listed specific properties or basic skills students should learn—first—to interpret expressions (e.g., the distributive property, PEMDAS). Others defined sensemaking more broadly, indicating that they would start by helping students understand how variables are used and behave.
- Teachers' strategies for moving students' thinking forward showed similar range. Teachers identified various ways to address misconceptions or areas of difficulty, ranging from class discussions to error analyses to using models or manipulatives. Science teachers suggested having students pose a question or hypothesis

then discuss with peers. A math teacher suggested "annotating verbal expressions" or underlining key words (a cross-discipline strategy emphasized at a number of schools). Another, again emphasizing sensemaking, suggested "deconstructing expressions" to help students understand terms or properties.

- In informal interviews, teachers also noted that "it would all depend…" on what other learning needs or styles they had observed in students, in where the problem or question posed to students came in a unit. In other words, context could matter a great deal.

These initial responses provided some process evidence of how teachers use PCK, and how we might more fully develop tools to measure it. They also presaged the challenges we might encounter in devising scoring mechanisms: all the diagnoses and solutions teachers offered were legitimate. We could note, descriptively, that some responses echoed new standards more than others, indicated newer vs. more traditional approaches, or reflected different degrees of concern about what students would need to succeed on high-stakes tests. But, there were few quantifiable differences.

In many ways, our initial foray into developing PCK measures reflected the findings of the 2012 Biological Sciences Curriculum Study PCK Summit, convened to discuss the complexities of defining and measuring PCK. One outcome of the summit was a chart of the "variety of knowledge domains" that influence PCK, all leading to a common desire to understand the impact of PCK on student achievement—the goal for MASTER as well. Their chart also portrays the interplay between teachers' knowledge bases, topic-specific knowledge, teacher and student beliefs, and classroom practice and context (Gess-Newsome 2015).

9.2.2 Larger-Scale Test with Mentors and Residents

After discussions with the MASTER PI team about findings from our small-scale test, we began development of a larger-scale set of tools designed to gauge residents' and mentors' PCK skills. The three-part assessment, which went through multiple reviews with the PI team to ensure content validity, asked mentors and residents to do what teachers routinely do: anticipate students' needs as they plan lessons (Part A), analyze student work to see what they do and don't know (Part B), and use that information for instructional planning (Part C). The activities were adapted from a variety of sources, including the Praxis 2, the National Assessment of Educational Progress (NAEP), and the Program for International Student Assessment (PISA).

Mentors completed two parts (B and C) of the online assessment, and residents, all three. An initial question asked teachers what subject they taught and then directed them to a subject-specific assessment for biology, chemistry, earth science, or math that began, in Part A, with questions about the big ideas students should understand. Parts B and C of each activity asked teachers to respond to examples of student work, gathered from various sources. The PI team advised against including a scoring rubric so teachers could see how they were being assessed, but each section included a "things to consider" section, based generally on existing scoring rubrics (Praxis 2, NAEP, or PISA rubrics, the edTPA rubrics recently adopted by New York, and the New Visions unit and lesson design rubrics).

9.2.3 Scoring

To develop our scoring methods, we reviewed these existing rubrics, alongside responses from residents and mentors from each subject area, looking for submissions that could serve as benchmark or anchor papers. Because the mentors did not complete Part A, we invited four to five local science and math teachers to complete Part A online, asking them not just to answer the individual items but also provide as many correct answers as possible; this generated, for example, a pool of "big ideas" that we could refer to as we scored responses.

Using existing rubrics and sample responses as guides, we drafted a two-part rubric. The first part rated responses to each subsection (A.1, A.2, B.1, B.2, etc.) on a four-point rubric (0, no or incomplete response; 1, beginning; 2, developing; 3, proficient; and 4, exceptional).

We used a holistic rather than analytical approach, assigning an overall score per subsection, based on an overall assessment of the response. When, for example, the question asked teachers to provide three big ideas or two to three strategies, we did not rate each one separately.

Each category gauged the extent to which teachers did what the question asked (e.g., "cite evidence from student work") and the overall substance of the answer. In the scoring, we tried to keep the MASTER approach in mind, focusing on big ideas, inquiry-based learning, learning progressions, and the degree to which the response focused on students' work, or what students did and didn't understand.

The second, more descriptive part of the rubric allowed us to code (rather than evaluate or rank) responses. The descriptive categories that made up the rubric emerged from our successive reviews of responses and included whether teachers focused on procedural knowledge or content/concepts, whether their strategies for addressing gaps in learning or misconceptions were teacher- or student-directed, and whether they reflected standards or practices, e.g., whether they suggested using or reinforcing vocabulary, modeling, or other ways to represent concepts.

This two-level scoring allowed us to assess or quantify PCK, and see if there were differences in responses from mentors and residents, or from new and more experienced teachers, and at the same time identify trends and account for some latitude in what constitutes good teaching.

Using an iterative process, we assigned scores and recorded descriptive data individually, reviewing scores to arrive at consensus, and adapting the rubric and scoring sheet accordingly. Two evaluators then scored the full set of math and science responses ($N = 21$ mentors, 20 residents), comparing scores to establish inter-rater reliability, but also averaging scores that were within one rubric score of each other. If there was a discrepancy of more than one, we discussed the scores and brought in a third scorer to reach consensus.

9.2.4 Results

Results showed some predictable differences between experienced and novice teachers and, again, variations within groups.

Overall, mentors scored a little higher than residents. Breakdowns by section and subjects also showed mentors scoring higher, though margins were fairly narrow, with means of 2.34 vs. 1.95 (on the four-point rubric) for the section on examining students' work and identifying what they did or didn't understand (Part B).

Residents scored highest on Part A ($M = 2.27$), which called for content knowledge, and lowest on C, which asked respondents to broadly outline lesson plans or instructional strategies for addressing the learning needs identified in student work. (See Fig. 9.1) Residents struggled some with scope, especially in what constituted a "big idea," math residents more so than science residents.

The descriptive scores revealed variations by discipline, even within subjects (i.e., biology, chemistry, and earth science) in the emphasis on procedural vs. content knowledge. Biology teachers, compared to those teaching chemistry or earth science, put more emphasis on students' content needs or deficits. Descriptive differences, however, did not provide the evidence we needed to validate our PCK assessment tools.

The results did provide a different kind of confirmation, however, one that echoed the kickoff findings: mentors, or experienced teachers, relied more on contextual cues. In some cases, mentors wrote that "they would need to know more" about the students to diagnose learning needs and design lessons around them. They talked more about individual learning arcs—how a student might think—than more generic, broadly applicable learning progressions. In focus groups and other informal conversations, mentors also stressed the need for residents, or novice teachers,

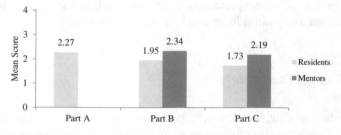

Fig. 9.1 PCK three-part assessment, overall PCK rubric score comparisons (Only residents completed Part A)

to have enough skills and strategies in their toolkits to help them think on their feet or address a learning need or misconception in the classroom, in the moment, rather than in the abstract.

9.3 Phase 2: A Narrower Focus—Embedding PCK Assessment in the Teaching Context

9.3.1 A Definition of PCK for MASTER

As the project moved forward in Year 2, the MASTER PI team saw the need to adjust their programming lens, narrowing the aperture to articulate a sharper image of what skills they hoped MASTER teachers would gain. They developed a set of PCK objectives that described what "teachers with a high level of PCK" do in the classroom to "communicate that knowledge to students."

This addition allowed us to increase our depth of field and triangulate data from other external evaluation activities, such as interviews, artifacts, and focus groups. Rather than focusing on a single metric to assess how teachers develop PCK, which was not generating interpretable data, this multi-method approach meant we could identify thematically similar findings across the data sources, confirm that our instruments and strategies were all measuring the same constructs—albeit perhaps with different degrees of depth or breadth—and establish convergent evidence.

Other research supported this approach. Participants in the 2012 PCK Summit (funded by NSF and the Spencer Foundation), for example, came to the consensus that PCK should be measured both indirectly through observations, interviews, and written reflections, and directly with questionnaires and pencil-and-paper tests (Gess-Newsome 2015). Other researchers had also triangulated observations, lesson plans, interviews, and written tests to construct cases of PCK development in an interdisciplinary science PD program (Smith and Liu 2014).

Perhaps most important, focusing on specific PCK objectives describing what teachers with a high level of PCK do *in the classroom* allowed us to embed our work more deeply in the teaching context. We were able, for example, to map our survey items about classroom practice onto the team's objectives and create crosswalks between the objectives, or specific constructs, and both teacher and student survey items to explore congruities in reports about classroom learning environments.

9.3.2 Lesson Plan Analysis

Findings from our PCK assessment had indicated that residents struggled more with designing lessons than experienced teachers. With that in mind, we began our Phase 2 battery of classroom-focused measures with a narrowly focused study of lesson planning, aligned to project's goals and PCK objectives. To see if the project's year-long training in lesson planning had an impact on residents' ability to diagnose student thinking, we analyzed a sample of their lesson plans from the fall, when residents were only a month into their clinical year, and the spring, when the year was coming to an end. We played off of two of the most salient PCK objectives laid out by the MASTER PI team: (1) using formative assessments to continually evaluate students' understanding of content, concepts, and practices and (2) employing various approaches to help students access all three.

Designing effective lessons is a key element of the MASTER residency model, as is the inquiry cycle approach that frames the lesson design process. As they move through the cycle, teachers diagnose students' learning needs using formative assessment strategies, plan and enact lessons around those needs, reassess, and reteach or reengage students to address lingering deficits. During weekly reflective seminars, residents examine student work and lessons with mentors. During a mid- and end-of-year Defense of Learning, they demonstrate that they can move students' learning forward by sharing findings that informed and confirmed their data-driven instruction. Lesson planning is also one of the six domains in the Danielson Framework, used not only in residents' evaluations but all NYCDOE teacher evaluations. The study design allowed us to leverage the fairly ubiquitous activity of lesson planning, and to access, through New Visions, samples of lesson plans.

To track the development of residents' PCK, we compared fall and spring lesson plans. We also cross-referenced elements of sample fall and spring lesson plans with MASTER's PCK learning objectives, looking for changes in the number of PCK objectives addressed in the lesson plans and the quality of those references. We expected that over time, teachers' lesson plans would contain more specific objectives, differentiated instructional strategies, and formative assessments. We also predicted that, in the spring lessons, teachers would ask more "higher level" questions to assess students' thinking and reasoning.

9.3.3 Overall Content Analysis

We analyzed a total of 14 fall/spring lesson plan pairs from six math and eight science residents. We first tracked the frequency of selected lesson elements using a coding scheme derived from the Danielson rubric, the MASTER PCK learning objectives, and the lessons themselves. We paid particular attention to the lessons' standards and goals, as well as the types of classroom assessment strategies teachers planned to use. Since the PCK learning objectives stress awareness of students' prior knowledge and misconceptions, we included those in our checklist as well. We calculated the percentage of lesson elements within the math and science, fall and spring samples, and compared frequencies at each time period using Fisher's exact test, a nonparametric technique similar to a chi square, but suitable for small samples (Leech et al. 2005).

We found that the fall lesson plans already contained a substantial amount of information about teachers' aims and activities. Almost all teachers listed aims and objectives and framed their lessons with assessments (i.e., a beginning Do Now and concluding Exit Ticket). Teachers typically articulated the questions they would ask in a whole-class discussion (100% and 87.5% of the math and science lessons, respectively), but less frequently shared strategies for assessing small groups (appearing in 25% of the math lessons and 56.6% of the science lessons). Lesson contents changed minimally from fall to spring, with no statistically significant differences across elements over time (all Fisher's exact values >0.05).

9.3.4   Coding Process

Once we had examined the overall content and structure of the lesson plans, we further explored the precise nature of the questions teachers planned to ask. We defined questions broadly to include any inquiries made in structured or unstructured contexts (e.g., a classwide Do Now vs. an informal small group discussion). We identified questions in the lesson plans themselves and in any attached handouts. Across the fall and spring lessons, we recorded a total of 185 questions in the math lesson plans, and 198 in science, with no statistically significant differences in the number of questions between teachers or time periods (math: $\chi^2 (5) = 5.81, p = 0.325$; science: $\chi^2 (7) = 10.01, p = 0.188$).

Two evaluators classified the cognitive depth of each question using the categories from Bloom's revised taxonomy of educational objectives (Krathwohl 2002). They received a list of questions in randomized order that had been cleaned of teacher names and the lesson time period (fall vs. spring). Raters did, however, have information about the section of the lesson in which the question appeared (e.g., direct instruction, activity, small group work), and a brief description of that lesson segment (e.g., "These are questions the teacher will ask if she sees students making errors on their lab handout") to provide some context.

The coding proceeded in two phases. In the first phase, we rated a 20–30% random sample of questions for each subject area and then discussed the ratings to clarify coding categories and reach consensus. During this phase, we realized it was difficult to reach exact agreement on the six Bloom's levels, especially because the category definitions overlapped (Krathwohl 2002). We decided to condense our rating levels to three: low (remember/understand), medium (apply/analyze), and high (evaluate/create). We also added a "cannot be determined" category for questions we could not rate without additional context.

In the second coding phase, we coded the remaining questions independently and then compared our ratings. We reached 76.8% exact agreement on the math questions, which is above the 70% threshold recommended for interrater reliability (Stemler and Tsai 2008). We reached 55.9% exact agreement on the science questions, because it was difficult to tell whether students had to apply information or simply recall it, and whether there was enough information to rate the questions.

Do Now questions, for instance, could serve as a review of previously taught information or an assessment of incoming prior knowledge, depending on the position of the lesson in the overall unit. Given these discrepancies, we reviewed and discussed each question individually to reach consensus and used these ratings in our analysis.

9.3.5   Results

Most questions fell into the first two levels of our coding scheme, reflecting the "remember, understand, apply, and analyze" categories of Bloom's revised taxonomy. On average, science teachers asked similar numbers of basic and more applied questions. They would, for instance, ask students lower-level recitation questions (e.g., "define kinetic energy") and then engage students in a lab activity requiring the interpretation of observations (e.g., "How can you account for this [the behavior of food dye in hot and cold water] in terms of the motion of the molecules of water?"). Math teachers tended to ask more medium-level questions than low, commonly demonstrating a mathematical practice or algorithm which students then applied to similar problems. While teachers in both subject areas asked application and analysis questions more often in the spring than in the fall, differences were not statistically significant (math: χ^2 (3) = 5.97, p = 0.113; science: χ^2 (3) = 4.30, p = 0.231). (See Fig. 9.2.)

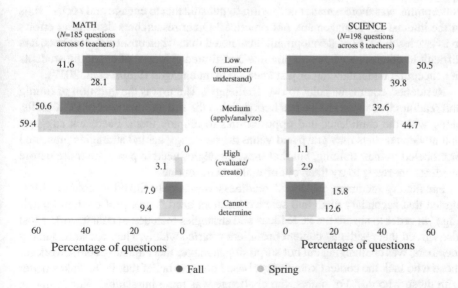

Fig. 9.2 Cognitive level of questions in residents' fall and spring lesson plans

Our analysis of lesson plans gave us additional insights about how novice teachers develop PCK—but in the end echoed conclusions from Phase 1 and our more abstract scenarios: context plays a major role in the PCK skills and strategies teachers bring to bear, and even narrowly focused metrics may still not be sensitive or time-bound enough to detect skills and changes over time.

9.3.6 Focus Group and Interview Feedback

To triangulate data embedded in the program context, we also used other data collection strategies to gather contextualized data, including focus groups and interviews. Conversations with MASTER residents, mentors, and faculty shed light on both the benefits and obstacles to folding PCK into practice and teacher preparation.

In focus group and survey feedback, MASTER residents—and graduates, in their induction years—repeatedly explained that, as novice teachers, they had trouble translating some PCK-focused training and skills into practice. They understood PCK conceptually, as a constellation of skills that effective teachers employ, but were uncertain how to operationalize those skills. Some residents noted that they "saw the importance" of PCK but planned to store ideas and insights away to draw on at a future date.

Others explained that they understood what they *should* do but were often consumed with "more mundane things," "day-to-day tasks," such as "getting through attendance or the Do Now." Reflecting at the end of the clinical year on attempts to apply PCK skills, one resident acknowledged that trying to draw out students' misconceptions was more a matter of "trying to get students to engage and focus," thus in the interest of "management, not practice." Other researchers describing efforts to assess teachers' PCK development have noted that, when observing teachers, it is difficult to determine whether a practice constituted evidence of engaging students in conceptual understanding or just keeping them engaged (Smith et al. 2017).

Residents' comments point to two challenges. The first is the question of timing and readiness, or when novice teachers develop the full complement of PCK skills, along with the confidence and opportunities to employ them. Feedback suggests that novice teachers may grasp and value the pedagogy behind strategies presented or modeled in their training, but that these strategies need to percolate more before residents are ready to try them out or make them routine.

For faculty members, residents' readiness was largely linked to content. Most agreed that secondary math and science teachers need "a grasp of content knowledge" in order to absorb the PCK ideas and strategies included in coursework. Most also agreed that residents' content proficiency varied widely. That, for some faculty members, was challenging but not surprising, a matter that adjusting coursework for those who lack the content knowledge, based on their belief that PCK "takes better with those who do." For others, the challenge was more intractable. Reflecting on efforts to design a course or assignment that fused pedagogy and content into the

PCK amalgam, one faculty member concluded, "I don't see how you can do it if you don't have a strong content foundation." Science faculty members, echoing mentors, alluded to content-heavy science courses, content that is "always changing," and the challenges of "pitching the course at the right level."

The second challenge is students' readiness. Residents and mentors alike were uncertain that high school students had the background for conceptual understanding. Residents especially were uncertain about how to fill in the gaps or use those gaps as a starting point for building conceptual understanding. One resident worried that trying to get at students' misconceptions and figure out how to help them "relate to a concept" could mean discovering that students have "no conceptions at all."

Some of the MASTER teachers' concerns stemmed not just from their experiences in translating PCK into classroom practice but also from making the shift to the NGSS and CCSS. Both sets of standards encourage a shift in teaching to big ideas, a larger grain size. That was the way we designed our PCK assessment: we canvassed project partners to find those cross-cutting concepts, something all teachers within a subject area would be familiar with, then, as the first step in the assessment exercise, asked teachers to frame the big ideas that students should learn. From there, they reviewed actual student work on an assignment related to the topic, looked for learning needs they could address, and mapped out instruction. The residents performed relatively well on the first part—they could articulate the big ideas—but seemed to falter when it came to crafting lessons. We were left wondering whether novice teachers simply lacked those skills, or whether our measures did not provide them with enough information and context to diagnose and target learning needs.

9.4 Phase 3: From Evaluation Challenges to Programmatic Opportunities—Viewing PCK as an Input Rather Than an Output

By the end of year 3, we had developed more contextualized measures but still had the problem of attributing variations to the MASTER program. This led us to further investigate the program elements in detail and take a retrospective look at what novice and experienced teachers need or want to learn about PCK and how partners might continue to integrate these into training, coursework, and professional development. In year 4, we thus focused more on our role as evaluators than researchers and the lessons learned from our challenges with measurement that might yield opportunities for program improvement.

What seemed most consistent in our findings were the variations within and across disciplines, which made it challenging to attribute changes to the MASTER model or make specific recommendations. The residents and mentors who were the participants in our studies were working not just in multiple disciplines but also teaching in different schools where curricular structures, unit plans, and pacing calendars could vary. It was thus challenging to identify a common denominator for what teachers were expected to do and learn around PCK. In other words, we couldn't study the topic-specific PCK that researchers say is the sweet spot for research and development (Gess-Newsome 2015).

Those who have been able to measure PCK with any degree of confidence have limited the scope of their studies, which has allowed them to elicit adequate variation in responses to detect consistent, interpretable patterns. Although we focused on specific aspects of MASTER—e.g., lesson planning—as program evaluators, we also had to look at the program as a whole, with all of its complexity and moving pieces, and didn't have the luxury of just isolating a single domain.

Our task did afford us a different kind of luxury: the multiple data sources we could assemble to make some holistic programmatic recommendations based on common threads and suggest some opportunities for further research. In our final interviews with faculty members and partners, we asked about their experiences teaching novice teachers about PCK, to see how their strategies and conclusions fit within established research frameworks. We interviewed residents concluding their clinical years and looking forward to their induction year, and MASTER graduates in their induction year, about what they had learned and how they might apply it going forward. In focus groups with mentors, who serve as an intermediary to this process, we asked them to reflect on what residents need in the longer term (i.e., a toolkit of resources), after they've gained proficiency in basic pedagogy and classroom management. Collectively, MASTER stakeholders shared three outstanding needs for teaching and learning PCK.

9.4.1 Finding the Right Grain Size

There was general agreement that the challenge central to coursework, resident, and mentor professional development, mentoring—and assessment efforts—was grain size. Deeper, more conceptual learning and a grasp of concepts and skills that cross or defy boundaries should certainly be the end goal of teaching, the place to start in backward planning a unit or lesson. The issue is how to chunk training for novice teachers. In response to an open-ended question on the end-of-year survey, a math resident suggested more training on "teaching topics" rather than "designing lessons." A smaller grain size—the topic-specific sweet spot that has helped focus PCK research—could also benefit teacher training.

9.4.2 Seeing PCK as Accumulated, Constructed Knowledge

A topic-specific approach is a starting point for a knowledge base that residents build on as they gain experience as teachers and stock their toolkits. In MASTER's final year, we asked mentors and residents where or in which aspect of training they gained PCK skills. What we found also supported the notion that PCK is constructed, or accumulated knowledge, and that some elements of training have a proper venue: a Hunter misconceptions math course, for example, gave residents the opportunity to share their own examples and experiences of student misconceptions and build a repository they could consult based on their experiences, those of others in the class, and those from the research. A science faculty member had directed residents in another course to the AAAS lists of misconceptions, listed by subject and topic. The goal is not to leave prospective teachers with the idea that content is always atomized, just to supply them with references they can consult as they frame their own lessons.

9.4.3   Surfacing Tacit Knowledge

For experienced teachers, the challenges involved in further developing PCK skills are a little different. Early on, when mentors were unfamiliar with the acronym, and even later, when it was more familiar, a common, oft-repeated response was, "that's what teachers do." Or "What else would we do?" other than figure out the best ways to teach the content in our subject areas to students, acknowledging what students bring to the topic, what framing they need, and how to figure out what they do or don't already know.

As other researchers have concluded (Smith et al. 2017), much of what teachers know about PCK may be "tacit knowledge." They know more than they typically articulate in pedagogical or research terms. The challenge, then, is how to bring it to the surface. The MASTER model already includes features and structures designed to encourage reflection: the inquiry cycle is built around deliberate, successive examinations of what students do and don't know. Mentors and residents routinely engage in reflective seminars devoted to analyzing practice. Mentors consider reflection as one of the most critical skills and habits a novice teacher can develop, a habit of mind that will serve them not just in their induction years but throughout their careers.

Mentor training focused on surfacing that knowledge could serve two purposes: it could strengthen mentors' metacognitive awareness of their own teaching practices and give them strategies for helping residents articulate and develop PCK skills. Analyzing student work together, as a collaborative exercise, would help teachers gain additional insights similar to those they derive from collaborative lesson planning, just focused, as the resident suggested, on teaching topics rather than planning lessons.

These discussions would draw on personal PCK—mentors' own experiences, contexts, subject matter—and confirm that PCK is not monolithic. At the same time, it might also surface, or help teachers collectively establish, some canonical PCK (see Settlage 2013). Certain tenets, agreed on by mentors in PD sessions, with faculty members in joint or collaborative sessions, could be valuable for residents. Much PCK for them will be personal, but a reference of canonical PCK might give them something solid to draw upon. Training could privilege both faculty expertise and mentors' expertise about what novice teachers need to know.

Perhaps partners should make their tacit PCK more explicit as well, and together reach consensus and begin to create a canon from which everyone can draw. That in turn might reduce some of the challenges for faculty members and mentors and variation in residents' experiences. Not that the goal should be standardization of the program, but rather greater fidelity of implementation, which could ensure that partners meet their goals and evaluation efforts generate evidence to confirm their success.

9.4.4 Conclusion

Our experience with MASTER reiterates the importance of evaluating PCK with teachers always in the foreground. Throughout our instrument development process, we gathered feedback from residents and mentors in order to construct highly contextualized tasks that reflected their ability to diagnose and respond to students' learning needs. Along the way, we began to pay increasing attention to the implementation of the MASTER training in order to interpret the variability we observed within and between subjects, data collection time points, and teachers with varying levels of expertise. We found that novice teachers may need more support assembling a collection of PCK practices—some canonical practices. They may not be prepared to use these as novice teachers, but, over time, they can draw upon this collective wisdom while also relying on practices forged from experience.

Program developers should continue to have teachers at all levels reflect on what they know about teaching and how they know it, in an effort to make tacit knowledge more explicit. This may lead to the construction of more sensitive, assessable scenarios that draw out teachers' knowledge for teaching specific science and math content, as well as the evidence used to support those practices. The dual challenge and opportunity for researchers and practitioners is to think about PCK in a way that facilitates group-level training but remains relevant, and true, to individual practices.

References

Ball, D. L., Thames, M. H., & Phelps, G. (2008). Content knowledge for teaching: What makes it special? *Journal of Teacher Education, 59*(5), 389–407.

Buschang, R. E., Chung, G. K. W. K., Delacruz, G. C., & Baker, E. L. (2012). *Validating measures of algebra teacher subject matter knowledge and pedagogical content knowledge* (CRESST Report 820). Los Angeles: University of California, National Center for Research on Evaluation, Standards, and Student Testing (CRESST).

Gess-Newsome, J. (2015). A model of teacher professional knowledge and skill including PCK: Results of the thinking from the PCK Summit. In A. Berry, J. Loughran, & P. J. Friedrichsen

(Eds.), *Re-examining pedagogical content knowledge in science education* (pp. 28–42). New York: Routledge.

Hill, H., Schilling, S. G., & Ball, D. L. (2004). Developing measures of teachers' mathematics knowledge for teaching. *The Elementary School Journal, 105*(1), 11–30.

Krathwohl, D. R. (2002). A revision of Bloom's Taxonomy: An overview. *Theory Into Practice, 41*(4), 212–218.

Leech, N. L., Barrett, K. C., & Morgan, G. A. (2005). *SPSS for intermediate statistics* (2nd ed.). Mahwah: Lawrence Erlbaum Associates.

Magnusson, S., Krajcik, J., & Borko, H. (1999). Nature, sources and development of pedagogical content knowledge for science teaching. In J. Gess-Newsome & N. G. Lederman (Eds.), *Examining pedagogical content knowledge* (pp. 95–132). Norwell: Kluwer Academic Publishers.

Settlage, J. (2013). On acknowledging PCK's shortcomings. *Journal of Science Teacher Education, 24*(1), 1–12.

Shulman, L. S. (1986). Those who understand: Knowledge growth in teaching. *Educational Researcher, 15*(2), 4–14.

Smith, P. S., & Banilower, E. R. (2006). *Measuring teachers' knowledge for teaching force and motion concepts*. Paper presented at the National Association for Research in Science Teaching (NARST) Annual International Conference, San Francisco.

Smith, E. L. & Liu, X. (2014). *The development of in-service science teachers' pedagogical content knowledge related to interdisciplinary science inquiry*. Paper presented at the National Association for Research in Science Teaching (NARST) Annual International Conference, Philadelphia.

Smith, P. S., Plumley, C., & Hayes, M. (2017). *Eliciting elementary teachers' PCK for the small particle model*. Paper presented at the National Association for Research in Science Teaching (NARST) Annual International Conference, San Antonio.

Stemler, S. M., & Tsai, J. (2008). Best practices in interrater reliability: Three common approaches. In J. W. Osbourne (Ed.), *Best practices in quantitative methods* (pp. 29–49). Thousand Oaks: Sage.

Wang, J., Schweig, J., Griffin, N., Baldanza, M., Rivera, N. M., & Hsu, V. (2013). *Inspiring Minds Through a Professional Alliance of Community Teachers (IMPACT): Evaluation results of the Cohort 1 math and science apprentice teachers*. CSE Technical Report 826. Los Angeles: National Center for Research on Evaluation, Standards, and Student Testing.

Part III
PCK in Informal Learning

Chapter 10
Pre-service Teachers Developing PCK in a Natural History Museum

Curtis Pyke, Tiffany-Rose Sikorski, Rebecca Bray, and Colleen Popson

Abstract This chapter is about the design and study of a pre-service teacher field experience that takes place at the Smithsonian National Museum of Natural History's immersive space Q?rius ("curious"). Q?rius features more than 6000 specimens including shells, skeletons, fossils, rocks, and minerals that are organized in non-text collections that visitors can see, touch, and study under a microscope. Prospective 7th–12th grade teachers from the George Washington University's Master of Education Program in Secondary Education were trained as museum volunteers to facilitate the Q?rius visitors' experiences and tasked with attending to visitor thinking, facilitating questioning, and sustaining engagement with the artifacts. Practical outcomes for science teaching resulted in gaining confidence working with learners, becoming better questioners, and learning to inquire with visitors (instead of giving answers to them).

Keywords PCK · Teacher education · Pre-service teacher · Informal learning · Inquiry · Science and engineering practices · Questioning · Natural history museum · NGSS · Museum collection · Secondary teacher education

10.1 Introduction

The real problem or challenge in our work with future teachers is to get them to "see" students' inquiry as the heart of teaching and learning from the start, providing them experiences that challenge their tendency to envision teaching only as exposing learners to canonical scientific ideas and concepts.

For over two decades, the science education community has honed its focus on a multidimensional view of science, identifying aspects of doing science that must be

C. Pyke (✉) · T.-R. Sikorski
The George Washington University, Graduate School of Education and Human Development, Department of Curriculum and Pedagogy, Washington, DC, USA

R. Bray · C. Popson
Smithsonian National Museum of Natural History, Washington, DC, USA

© Springer International Publishing AG, part of Springer Nature 2018   177
S. M. Uzzo et al. (eds.), *Pedagogical Content Knowledge in STEM*,
Advances in STEM Education, https://doi.org/10.1007/978-3-319-97475-0_10

developed along with what has traditionally been considered *content knowledge* (Duschl 2008). Major efforts to expand the aim of science education include the American Association for the Advancement of Science Project 2061's Benchmarks for Science Literacy *Habits of Mind* (AAAS 1993) and the National Science Education Standards' (NSES) focus on *Inquiry* (NRC 1996; CSMEE 2000). These prior efforts have directly influenced today's Next Generation Science Standards (NGSS), a complete reorganizing of the subject along three dimensions: *core disciplinary ideas*, *science and engineering practices*, and *crosscutting concepts* (National Research Council 2012; NGSS Lead States 2013). This shift in science learning outcomes for schools challenges us, as science teacher educators, to adapt teacher education curriculum and instruction to cultivate a more balanced and dynamic vision of school science among future science teachers. To that end, this chapter focuses on the potential of museum experiences as sites for developing inquiry aspects of *pedagogical content knowledge* (PCK) (Shulman 1986; Wood 2003; Davis and Krajcik 2005; Gess-Newsome 2015; Hill et al. 2008). Developing PCK is recognized as one of the three critical areas where today's science teachers need to develop greater expertise in teaching toward NGSS (NASEM 2015), and we believe PCK targeting inquiry and practices is worth specific attention.

What follows describes the conceptualization, development, piloting, and study of a field experience where pre-service teachers worked directly with visitors to facilitate exploration of artifacts in the Q?rius ("curious") space at the Smithsonian National Museum of Natural History (SNMNH). A team consisting of math and science educators from George Washington University (GWU) and science educators at the SNMNH collaborated in the design of the experiences supported by a grant from the National Science Foundation (DUE 1439819).

10.2   Developing PCK for Science: A Focus on Students and Inquiry

A young boy came with his mother, and he seemed to be so much interested in rocks. I didn't have much knowledge on rocks, so I just observed how they interact with rock specimens. The boy's mother asked questions such as "How does the rock feel? What colors can you see from those rocks? What can you see if you see under the microscope?" and so on. He was only 6 to 7 years old, and during an hour or so at the collection zone looked so much productive. If I look back to when I am with my daughter, I do those things too; I question her at the level of her knowledge, try to make her think for herself to find out noticeable things, wait for her rather than just giving off facts that I know. It doesn't matter how much I know, I just do my best to help her learn by herself and improve her critical thinking skills. Then, why is it so difficult to do the same thing with other kids? This is my question to answer to be a better volunteer at Q?rius, and a teacher in near future.
  –Teacher candidate journal

What does PCK for science look like in the age of NGSS? The construct of PCK evolved from an interest in teachers' professional knowledge and skills, highlighting the interactions between teachers' subject matter knowledge and general pedagogical knowledge (Shulman, 1986). Historically, PCK included representations and tools for teaching content knowledge (i.e., disciplinary core ideas) as well as knowing students as learners of content knowledge. However, as the notion of content knowledge in science grows into multiple dimensions, including science and engineering practices, so too must the notion of PCK. Consistent with the idea of PCK for disciplinary practices (Davis and Krajcik 2005), we try to distinguish here the PCK of knowing students as learners of subject matter ideas (i.e., learners of core disciplinary ideas) from knowing students as learners of science inquiry (e.g., learners of practices) while acknowledging that the two are inextricably linked in theory and practice (NRC 2012). In this chapter, we foreground PCK for disciplinary practices, presenting evidence of the emergence of pre-service teachers' knowledge of students' inquiry and the approaches that they can use to teach inquiry.

10.2.1   Knowing the Development of Students as Inquirers

The idea of science as inquiry and therefore students as developing inquirers was reified with the publication of the NSES. Grounded in research of that time (circa 1996), teacher educators had evidence that students in the early grades could learn to investigate earth materials, organisms, and properties of common objects, ask questions, construct reasonable explanations for observations, and communicate about explanations. However, they had trouble developing the logic of explanations supported by evidence. By the middle grades, it was believed that students as inquirers could learn to connect evidence and explanations, know that knowledge and theories guide scientific questions and investigations, design simple investigations and experiments to test ideas, and produce oral and written reports of their inquiry and its results. However, many had trouble understanding complex investigations where multiple variables influence an experiment. There was evidence that at the conclusion of high school, some students learned all of the above and also developed an appreciation of the logic of theory-driven experiments to test ideas. However, some still struggled with multiple variables in experiments, had trouble dealing with anomalous data, and struggled to reconcile their prior knowledge and experience with evidence from experiments about the natural world (NRC 2000; 1996). Also emerging from this early work was an assertion that inquiry is domain-specific, with inquiry procedures and strategies in science, technology, engineering, and mathematics all unique to each discipline. For example, the application of a general principle often occurs during science inquiry, and this strategy was considered particular to the discipline of science (NRC 2000).

Today, inquiry is constructed in ways less indexed by age and conceived as more of a complex progression of concepts, practices, and meta-level processes (Kuhn and Pease 2008). These researchers, along with others that have continued to study inquiry among students, provide evidence that inquiry is teachable and that inquiry-oriented curriculum has a positive effect on early adolescents (Furtak et al. 2012). Kuhn and Pease (2008) show the challenges and possibilities of targeting inquiry over several years and conclude that the development of inquiry proceeds across general categories including formulating objectives for inquiry, interpreting evidence and drawing conclusions, predicting outcomes of multivariate experiments, and representing and communicating findings. What this work makes clear is that inquiry can develop through multiple, planned experiences over time.

10.2.2 Knowing Approaches for Teaching Inquiry

If we view science as inquiry (Schwab 1958), then knowing how to teach science means knowing how to teach inquiry. Teachers prepared with approaches to teach inquiry and ideas about how inquiry develops have the foundation for making good decisions concerning when and how to support students' meaning-making (Mortimer and Scott 2003).

To support inquiry, teachers need to be versed in at least three kinds of strategies: (1) sparking, supporting, and focusing student inquiries on generative phenomena, (2) orchestrating discourse among students about scientific ideas, and (3) encouraging and modeling particular skills during inquiry (NRC 1996, 2000, 2007). The first area reflects what teachers do to use their knowledge of core disciplinary ideas and their knowledge of students as learners to guide inquiry (NRC 2007). The second area prioritizes strategies for eliciting, noticing, and responding to student ideas during the discussion, as well as orchestrating peer-to-peer discourse (Robertson et al. 2015). Such strategies emphasize support for argumentation and social moderation in the classroom and the laboratory. For example, instructional models such as argument-driven inquiry teach teachers to engage students in scientific argumentation within the context of core ideas (Grooms et al. 2015). Arguably, the first two categories of teaching strategies require techniques for doing inquiry with students; however, explicit attention to modeling inquiry in its fullest sense for the long-term development of inquiry itself must also be a focus of teaching. Theoretically, science inquiry is a type of generative process for learning (Mayer 2011). As such, general strategies for fostering generative processing such as questioning, breaking down problems into parts, showing how to accomplish tasks with explanations, asking the learner to explain ideas and procedures, and providing appropriate and timely feedback all should apply to advancing learners' capacities for inquiry. As the NGSS science and engineering practices mirror these instructional techniques, it supports arguments for the dual nature of inquiry in science education, being at once an outcome and a pedagogical approach.

10.2.3 Focusing on Experiences to Promote Teachers' Use of Inquiry

We know that simply telling future teachers about inquiry development and approaches to inquiry is not sufficient for establishing inquiry in teaching practice. Further, learning to teach with and for inquiry can be difficult because personal disciplinary subject matter knowledge, including perceptions about the nature of science, interact with teacher education experiences (Coffey and Edwards 2015). The NRC (2000), noting such concerns, suggested that teacher education experiences needed to adapt to be more learner-centered (i.e., considering the teacher as a learner), knowledge-centered (i.e., including subject matter and PCK), and assessment-centered (i.e., using more authentic practice with feedback). More recently, the National Academies of Sciences, Engineering, and Medicine (2015) echoed this call for more and different experiences in science teacher learning, specifically mentioning that teacher education programs must allow for (a) teachers to focus on specific subject matter and how students engage with that subject matter; (b) actively engage teachers in analysis of student work, observations of colleagues, and trying out and reflecting on new approaches; (c) collaboration among educators who share district, school, and/or department contexts; (d) sufficient duration to allow repeated practice and reflection on experiences; and (e) opportunities for teachers to seek coherence between school, district, and state policies and priorities. The principles from these consensus documents suggest that teaching science as inquiry might develop through experiences that engage future teachers with students who themselves are engaged in authentic inquiry. In addition, teachers may need specific training and modeling of approaches, practice implementing those approaches, and time to reflect on their effects.

In light of these ideas, we have considered out-of-school educational environments, where inquiry is bounded and encouraged, as ideal places for providing access to the kinds of experiences future teachers need as they develop inquiry teaching (Stein and Rankin 1998; Chin 2004; Jung and Tonso 2006; Aquino et al. 2010). Notably, museums provide future teachers repeated opportunities to interact with learners, in spaces and with artifacts that invite inquiry (Adams and Gupta 2017). Further, by participating in active museum work, future teachers can engage in authentic inquiry with their attention focused specifically on visitor/student behaviors as they engage with artifacts. Such experience, when tied to a university-based teacher preparation program, can support teacher candidate's training in, practice with, and reflection on specific strategies for facilitating inquiry (Rivera Maulucci and Brotman 2010).

10.3   Teaching Field Experience at the Smithsonian National Museum of Natural History

Motivated by the possibilities of museum spaces for learning to teach science as inquiry, we designed a fieldwork curriculum for pre-service teachers around the affordances of the SNMNH visitor/student experience. The SNMNH maintains a force of over 200 volunteers that support all aspects of the museum, including administrative tasks, exhibit and collection maintenance and design, and visitor experiences. The museum contains multiple immersive spaces (e.g., Q?rius, Butterfly Pavilion, Insect Zoo) where volunteers work directly with visitors. We initially selected Q?rius as appropriate for pre-service teacher candidates because of the educational mission, that it attracts school-age children, and because it allows for a variety of interactive volunteer-visitor inquiry experiences.

Q?rius is intentionally designed, with unique affordances for visitor experiences and teacher candidate learning. Q?rius is a vast space within the museum, distinct from the other exhibit spaces in that it is designed specifically for preteens to interact with real artifacts from the museum collections. There are lab tables and microscopes throughout, and the wall units and room dividers are filled with drawer after drawer of artifacts that visitors study by viewing, touching, comparing, and measuring. It is a library where instead of books, visitors find more than 6000 specimens including shells, skeletons, fossils, rocks, and minerals. These artifacts are professionally organized and made available in non-text collections, where the learner chooses the starting point as well as the pathway for investigation and where specific research areas are implicitly represented in the space (e.g., anthropology, botany, entomology, invertebrate zoology, paleobiology, vertebrate zoology, and mineral sciences). Visitors pursue their interests by viewing, touching, and looking close with the digital microscope. They can discover what looks like hair on the leg of a cool bug or feel the shape and surface of a new favorite crystal. They can also record images and notes in their own personal digital field book and use other tools in the space to find answers to questions. Visitors are not alone in Q?rius as there are sometimes real scientists and teachers there to promote noticing things about the objects, demonstrate different ways to look at them, or provide help finding related artifacts. All of these elements are designed to spark visitor inquiry and create pathways for deeper engagement with the collections.

10.3.1 Design of the Summer Q?rius Field Experience

The initial development of the Q?rius fieldwork curriculum was a component of a capacity building project with collaboration and sustainability goals. For the curriculum to be sustainable, we knew we had to make use of the existing training and mentorship structures within both the SNMNH and GWU's secondary education M.Ed. program. We hypothesized that placing pre-service teachers in museum volunteer positions could simultaneously contribute to the SNMNH volunteer corps and fulfill the community-engaged teaching (CET) field hours required as part of GWU's secondary education program. A review of museum goals and GW program goals suggested a framework for the experience. Figure 1 shows the alignment of goals and the resulting fieldwork/internship goals that were initially targeted. Further, the desire to not compete or interfere with the customary school-based fieldwork of the M.Ed. program resulted in focusing this effort on a summer field experience option. This approach has been successful to date; three cohorts of teacher candidates have completed the Q?rius summer field experience, supported by ongoing collaboration between educators from SNMNH and GWU.

Because in this model the pre-service teacher candidates are hired as official museum volunteers for the summer, they complete all of the registration, orientation, and training that a regular museum volunteer completes. First, they register using the Smithsonian Volunteer Portal and complete their federal background checks. Second, the pre-service teachers attend a 1-day orientation and 1- or 2-day training for their specific exhibit along with all other new summer volunteers at the museum. The orientation addresses (i) discovering behind-the-scenes scientific research, (ii) exploring activity content as learners, (iii) understanding the museum education approach, (iv) working with multigenerational visitors, and (v) examining scientist communication techniques.

In addition to the general training, teacher candidates complete a day of specialized training for the Q?rius Collections Zone. Teacher candidates in the Q?rius Collections Zone learn about collection-based inquiry (Sunderland et al. 2012), explore the collection themselves as learners, practice facilitating learner interactions with other volunteers, and shadow an experienced volunteer working with visitors. In total, pre-service teachers receive a minimum of 16 h of specialized training from museum staff as part of the summer field experience.

After training, the pre-service teachers begin working directly with visitors in their assigned spaces. Volunteer shifts at the museum run from 9:45 am to 1:45 pm or 1:15 pm to 5:15 pm daily from June 15 to August 26. Using the online museum volunteer scheduling system, teacher candidates sign up for the 4-h morning or afternoon shifts. Each shift begins with a 15-min volunteer check-in, where they learn about important events happening at the museum that day. Teacher candidates are required to complete ten 4-h shifts for a total of 40 h in the museum. Ongoing involvement with the museum is encouraged, and a few participants stay on as museum volunteers after meeting the minimum requirement.

The teacher candidates also participate in four 1-h seminars hosted by the GWU Secondary Education program to support the summer CET field experiences across the program. The first session is an orientation where teacher candidates are introduced to program and project staff and the components of the field experience. Later workshops build upon the questioning and engagement strategies practiced. These sessions also allow teacher candidates to share what they've been learning in their interactions with learners and to make connections to their own future classroom teaching. When possible, Q?rius educators participate in these sessions. Special topics related to Q?rius include recording discoveries in digital field books, asking generative follow-up questions, and sustaining visitor engagement.

The teacher candidates keep digital journals, writing weekly about their interactions with museum visitors and volunteers. At the end of the summer, GWU and SNMNH educators review these journals independently, identifying recurring themes and interesting anomalies that speak to successes and challenges that teacher candidates experienced at the museum. Following the independent review, GWU and museum staff meet together to discuss the journals, reflect upon the previous summer's successes and challenges, and identify any potential changes to the field placement experience for future summers.

10.3.2   The Pre-service Teachers and Their Experience as Volunteers

Here we describe the experience of two small cohorts of pre-service teachers, five over the first 2 years that volunteered for the Q?rius experience as a means to fulfill their CET fieldwork requirement for the M.Ed. program. These pre-service teacher volunteers varied in age, content expertise, and education experience. For some this was their first experience in teaching science and their first experience with the SNMNH, while others had some teaching experience and familiarity with the museum space. Most, but not all, were pursuing a science or math teaching license.

As planned, the experience began for the pre-service teachers as they entered into a formal relationship with the museum as volunteers. They participated in their volunteer training, shadowed experienced volunteers, observed school visits, and eventually lead visitors in engaging the artifacts. Our holistic impression at the start was that most teacher candidates brought an initial concept focus rather than an inquiry curricular aim to their experience, conceiving of the space a place for visitors to search for knowledge and for volunteers to answer visitor questions. Consistent with this focus is that some expressed concern about the adequacy of their content knowledge relative to the content of the space. Most of the pre-service teachers' concerns changed as they became more comfortable and confident interacting with the visitors and the artifacts. In general, we saw changes in the pre-service teacher experiences over time similar to what Kelly (2000) found in work with elementary pre-service teachers in museums. The teachers seem to move from initial concerns about their content knowledge and answering questions to more of a collaborative inquiry with the visitors and to greater awareness of the visitors' needs and interests.

The following comments were compiled from various participants to provide a sense of some benchmarks in the flow of the experience from beginning to end. On the first day:

> When I first started Q?rius volunteering, I was shy and afraid to talk to any visitor. I just opened boxes [when asked to] and responded to questions.

After even one or two sessions, the pre-service teachers were engaged and attentive to the visitors' experience:

> My second shift with the Forensic Mystery case was very fun as well. Visitors had fun too, but children under 10 did not understand…

After several sessions they gained confidence in supporting visitor interactions:

> After the first couple of hours the lab became so busy that I had to facilitate…I was able to help the visitors by asking questions and exchanging ideas. I became so much confident that I did have most of the knowledge that was required to ask and answer.

At different points in the summer, with support and encouragement, they all practiced and discussed different questioning techniques:

> …my observer advised me to try and open with a visual question, so that the awckward [sic] transition from question to lecture can be avoided. Taking her advice, I changed my opening question…and asked them "Can you tell me what you can see in this rock?…

And as the experience progressed, they could contextualize the experience in terms of their own growth and development:

...after 32 hours Q?rius made me more confident to ask and facilitate visitors to think as well as keeping me to think and seek out more knowledge. I know this is only a start. I have to improve more and more until I can be like numerous other experienced volunteers...and be a good teacher.

The 16 h of training plus the 40-h service experience seemed just enough to develop expectations for visitor interactions, begin to practice some skills, and to establish some anchor experiences for analyzing and reflecting on the visitor/learner experience. However, we do wonder if a longer experience might allow for additional instructional opportunities related to inquiry-oriented PCK as well as other NGSS practices to be targeted, not just questioning. As a 40-h experience, the pre-service teachers all expressed that the experience was worthwhile and that they had learned from it.

10.4   What Pre-service Teachers Learned at Q?rius

Exploratory research was conducted to develop an initial description of what was happening during the fieldwork and to assess the secondary pre-service teachers' progress toward the internship goals of facilitating questioning and sustaining engagement (see Fig. 10.1). The research was designed to collect and analyze qualitative data from the participants. Four volunteers agreed to participate and completed pre-experience surveys, weekly journal entries, and postexperience interviews as part of the research. The journal prompts asked pre-service teachers to reflect on learners' ideas and interactions; what visitors said, did, asked, and explored in the space; and what patterns emerged from observing learners' interaction with objects and/or the space. The data were analyzed first by developing holistic descriptions of the interns' experiences (see prior section) and second, by focusing on intern learning. The search for learning statements revealed notable references to several of the events specifically planned for the fieldwork experience. The events that emerged as salient to the interns included the training sessions and seminars, participation in planned activities (e.g., the forensic mystery case), shadowing experienced volunteers/educators, observing visitors (e.g., a mother and son), facilitating with visitors, feedback from observers, and journaling. These events became initial units for partitioning learning statements. We looked more closely within and between these units specifically for evidence of learning (a) about student/visitor inquiry and (b) strategies for instigating and sustaining inquiry. The two sections that follow present a sketch of our initial ideas about the learning of these pre-service teachers in these areas.

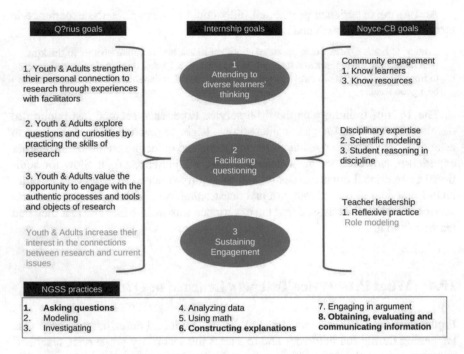

Fig. 10.1 Goals for teacher candidate learning in Q?rius. (Figure created by Estelle Raimondo, Project Evaluator)

10.4.1 Sustaining Visitor Engagement in Inquiry

The journal data, in particular, show two categories of comments emerging about the nature of visitor inquiry, observations about *visitors' engagement*, and *explanations for the engagement*. In the first category, engagement in inquiry, the pre-service teachers noted both similarities and differences in how different age groups responded to activities. In a number of the reflections, they noted consistency across age groups, "young and old first explored by curiously staring at the specimens." While in noticing differences, it was somewhat common to note engagement of just the younger visitors, "children under ten did not understand." The pre-service teachers also made reference to how the visitors interacted with objects often by first inspecting with their eyes or hands. They also noted changes in interaction or engagement over time usually from more generalized hands-off observation to more hands-on interaction with questioning. One pre-service teacher also noticed a difference in the quality of engagement by noting when one pair of visitors, described as more advanced, were observed testing their knowledge by creating a game around features of the fossils and guessing at what they were studying prior to scanning the tag.

The second category of observations concerns the pre-service teachers' explanations for engagement they observed. More than one pre-service teacher suggested that "learning styles" might explain when "visual" versus "tactile" learning strategies were used. Another expressed surprisingly that "even in this hands-on activity, students learn differently." Comments like these were taken as evidence that pre-service teachers were bringing some ideas about learning to their observations and using these ideas to try to explain instances of visitor engagement. A couple of comments suggested that engagement may also be influenced by some level of expertise. An example of this is the discussion noted above of a pair of visitors systematically engaging a collection of artifacts with a game-like approach. This pair was described by the pre-service teacher as "experts [in the content area of the fossils]" presumably in contrast to the more naive approaches of other visitors observed. The pre-service teachers also attempted to explain the different types of visitor engagement by ascribing motives or motivation to their inquiry. For example, visitor actions were identified as driven by either surface-level curiosity such as "just wanting answers" or more of a mastery orientation indicated when believing some visitors needed to "know everything." In other cases, they took the time to note and comment on when visitors appeared excited, engaged, or were just having fun. They explained this through the inherent interestingness of the artifacts (e.g., pyrite looks like real gold), or an interest in using the tools of the space (i.e., microscopes), or by idiosyncratic personal factors. That qualitative differences in approaches to inquiry were noted, with attempts at explanation, suggesting teacher candidates were developing models of inquiry growth and development.

What is promising here is that some of the observations map to what we hope pre-service teachers might observe and learn about what might occur during an authentic inquiry, such as observing, touching, asking questions, making and checking hypotheses about classification, sorting, etc. It is interesting to us that their ideas about student inquiry seem intertwined with their ideas about learning. In retrospect, it is somewhat intuitive to try to apply learning theory if one assumes the visitors are trying to acquire knowledge and learn about things. However, there seems to be significant potential here to work on developing contemporary models of learning and to explore the connection between science inquiry and learning. More remains to be understood about how the pre-service teachers are making sense of the dual nature of inquiry: as science and as a pedagogical approach.

These data also highlight for us what teachers might learn from observing visitors of different ages engaging in inquiry. New teachers may underestimate the abilities of their students, particularly if they misapply concepts of Piagetian development to assume that young learners are only capable of engaging in "concrete" experiences in visual or tactile ways (Metz 1995). Yet, in Q?rius, visitors young and old can be observed approaching the collection in similar ways.

10.4.2 Facilitating Questioning

An underlying theme of the data is the utility of the experience for building efficacy for questioning strategies that facilitate visitor engagement. This outcome was highlighted in the program evaluation where it was noted that the teacher candidates all developed appreciation for and skills in asking questions to support inquiry. As expected and intended, Q?rius clearly offered authentic and challenging moments to interact with visitors around questioning. A number of reflections suggested practice or growth in questioning, resulting from being pressed into question asking and answering, either through training or as a natural result of the volunteer experience. Perhaps because of the press toward engaging in question-answer discourse, most of the pre-service teachers felt they got better at questioning and for them better meant realizing they did not need to know answers to all visitor questions and realizing they could respond to visitor questions in ways other than giving an answer, using different types of questions, using open-ended questions, learning to ask why questions, and encouraging attention to detail with questions. Interestingly there is also evidence that teacher candidates reflected on questioning practices that did not work. For example, one candidate said:

> I am able to ask some (but not diverse) questions, but most people with me just respond by saying I don't know…someday I can be like [expert volunteer] making people more interested just by asking questions.

And another noted that asking a question alone does not work without more follow-up and support in helping the learner to investigate possible responses.

The data also show agency in remedying less than successful questioning. The pre-service teacher volunteers were quite resourceful in finding different approaches as they recalled the skills of those they shadowed, noted parents' behaviors with their own children, and how other volunteers engaged visitors with questions. They also began to differentiate and notice the various types of visitor questions as bids for different kinds of responses. They identified questions that seemed to demand quick yes/no answers, others as curious why questions, and questions that asked how that invited more exploration.

Most pre-service teachers reflected on improvement in questioning specifically attributed to (a) having the opportunity to practice at the museum, (b) noticing productive types of questions for particular objects (e.g., in the case of rocks, "How does it feel?"), (c) improving content knowledge such as learning more about skulls and bones to enable better questioning, and (d) explicit training in questioning strategies to use at the museum (i.e., workshops, shadowing, seminars). This finding is encouraging because effective questioning promotes generative processing and learning (Mayer 2011) and is part of good teaching. More pragmatically, questioning is also one of the eight NGSS practices that compose science inquiry, and we believe that modeling good questioning as part of doing inquiry with students is a means to promote inquiry development.

10.4.3   Teaching Science

While studying our data for clues about the relevance of these experiences for pre-service teachers learning to facilitate questioning and sustain engagement, two other noteworthy themes emerged. One theme related to the ongoing salience of the subject matter embodied by the collections and another suggesting the experience was serving as some type of preliminary test of teaching ability. Both reflect the pragmatic orientation these candidates brought to their teacher preparation at the masters level.

Not too surprising was that the subject matter of the collection was at times overshadowing or at least being reconciled with the press to attend to the visitors' engagement and interaction with the artifacts in the collection. For one participant the subject matter (e.g., anatomy) clearly took center stage when it was implied that inquiry "involving the human skeleton" might play a supporting role toward learning the content, "as part of my launching lesson." Another pre-service teacher expressed concern that if some subject matter was not learned, the visitors might not take anything away except merely a liking of the interactivity of the space. It is hard to say for sure from our small sample and without some follow-up, but it seems possible that at least a couple of the pre-service teachers maintained a subject matter orientation throughout the experience. Certainly, there is value for the visitor who, as a result of interacting with the teacher volunteers, develops new understandings about the artifacts in the space. However, the teachers' sense that the interactions were *only* valuable if the visitor leaves with factual knowledge about artifacts, aligning concerns that teacher candidates' epistemological stances may confound the teaching and learning of inquiry strategies, especially toward the end of developing inquiry itself.

The second theme that emerged tangentially to our focus on inquiry was that the pre-service teachers were somewhat spontaneously making sense of their experience relative to their ultimate goal of becoming teachers. They made self-evaluation comments that suggested the experience made them aware of who they were, or were not, relative to others they perceived as good teachers, often the more experienced volunteers. The experience caused one to become concerned about their expertise in science. Asked directly by a visitor, "what are you an expert in?" the pre-service teacher seemed to worry about not having an answer and questioned if teachers ought to be an expert in something. Another expressed some frustration communicating with the visitors noting that this was something they were going to have to work on to be a teacher. There is also a set of comments where the pre-service teachers tried to explain or qualify their performance as new, or their first time, or just a start. Almost always these comments were followed by an assessment of the need to improve to become a good teacher. At this point, we wonder if our planned curriculum, including the observation feedback, required journals, seminars, and the research focus drew out these assessments or if there is something about the work in Q?rius that stimulates a natural move to self-reflection and improvement.

10.5   Discussion and Recommendations for Future Work

PCK for teaching inquiry is the foundation for ambitious science teaching (Thompson et al. 2013) and necessarily includes knowing students as inquirers or doers of science (Duckworth 1996; Levin et al. 2012). Thus, the development of future teachers' PCK for inquiry-oriented science ought to enable them to both effectively support their students in doing science to learn core ideas and to use doing science to teach inquiry itself. Yet, we are keenly aware that future teachers may not automatically be ready to take up PCK for inquiry from passive exposure to students' inquiry experiences because they themselves may not have experienced practices and inquiry as described in NGSS (NASEM 2015). Their orientations to the discipline (Coffey and Edwards 2015) and subject matter knowledge (Hill et al. 2008) influence what they attend to when engaging youth in inquiry.

While the development of core disciplinary ideas is viewed as inextricably linked to students' engagement in inquiry and science practices (NRC 2012), PCK for inquiry and science practices (Davis and Krajcik 2005) remains tacit knowledge that seems to emerge over time through teachers' first-hand experiences guiding and encouraging inquiry with learners. Further, many university-based programs offer few opportunities for engaging pre-service teachers in sufficient authentic student inquiry experiences for the purpose of teaching them about the how-to of science as inquiry, and most exposure is reserved for the final student teaching phase of a program. The professional education community, reacting to the need for more inquiry-oriented teaching, critiques the academic preparation of science teachers and argues for policy to support special courses and new experiences for teachers' in-depth study of their subjects (NRC 2000; NASEM 2015). In the work presented here, we have looked to a museum-based field experience for its potential in building on existing disciplinary expertise by providing pre-service teachers training in facilitating inquiry early in their program of study.

Museums, community centers, and historic sites have great potential for teacher learning (Avraamidou 2014; Seligmann 2014). In the case of Q?rius, we were drawn to the unique affordances of the space and materials for promoting visitor choice, curiosity, and interaction. Design elements such as the arrangement of the artifacts, content of the artifact labels, presence of digital microscopes and field books, and the presence of trained volunteers who interact directly with visitors create an experience where concepts are embodied, where cross-cutting themes can be made visible, and where practices can unfold all somewhat natural for visitors. The structure of the spaces allow pre-service teachers a chance to participate in a rich, authentic, and reliable experiences facilitating inquiry learning with diverse learners.

To foster the development of pre-service teachers in the museum space, GWU education faculty and SNMNH science educators collaborated to create a curriculum grounded in a 40-h volunteer experience that targeted facilitation of inquiry for the museum visitors. The specific goals for the pre-service teachers include (a) attending to diverse learners' thinking, (b) facilitating questioning, and (c) sustaining engagement. The larger grain educational objectives of the project were met as the pre-service teachers successfully completed their training and effectively served as volunteers in the space. This chapter provides some evidence of accomplishment around the smaller-grain size teacher learning goals.

When we looked closely at what the pre-service teachers attend to concerning visitor thinking and engagement, we see potential groundwork for future learning. We found evidence that the pre-service teachers observed behaviors and patterns useful for advancing and teaching inquiry, and we wonder how best to make use of their observations in their future coursework. We are encouraged that the experience generated first-hand knowledge of inquirers, and we found it interesting that they applied their ideas about learning to explain patterns of engagement. This suggests to us that the experience also creates an opportunity to explore models of learning that make a case for teaching science as inquiry (Smetana et al. 2017). We also think that their observations and explanations open the door to ask what is not being observed in terms of the eight NGGS practices and why.

We see much potential in teacher learning when considering specific aspects of inquiry that we encouraged and that were observed. For example, with questioning, one of eight NGSS core science practices, we have evidence of pre-service teachers noticing and responding to visitor questions and learning to ask questions (i.e., model questioning) to initiate or move an inquiry forward. Almost all pre-service teachers practiced questioning, reflected on their competence, and articulated a desire to improve in this capacity. This may have been out of necessity or due to the goals set by the team. In either case, the questioning experience seems a fruitful place in the future to make explicit notions of PCK for teaching inquiry. We wonder if explicit attention to other NGSS practices such as investigating and constructing explanations might also generate worthwhile anchor experiences.

However, we find that engaging in inquiry with visitors may not be intuitive for all pre-service teachers at this very early phase of their teaching development. It is likely that more needs to be done to make these experiences a catalyst for teaching science as inquiry, especially for those that bring a strong science as concepts orientation to their work.

We leave this work with excitement about the potential of long-term collaboration and to ongoing refinement of the curriculum to focus more on the PCK of disciplinary practices. We remain convinced that teacher learning can and must occur outside, as well as inside, of schools and that it makes sense to seek out places for teachers to learn where science learning is happening with a high frequency outside the classroom. While we recognize that the evidence is limited as to the best practices for employing out of school contexts for teacher learning and that the data here are limited in generalizability, we hope the methods and experiences described here contribute to developing a knowledge base.

This work has specifically stimulated our interest in the future to:

- Expand to other spaces in SNMNH such as the Insect Zoo and Butterfly Pavilion and to study the effects of different affordances of spaces on pre-service teacher experiences.
- Help the pre-service teachers build on their efficacy for establishing interest and engagement with the artifacts to facilitate deeper inquiry.

- Establish more emphasis on inquiry development, and introduce more practices such as investigating and constructing explanations.
- Understand more about the visitor experience so that we can better scaffold teacher learning around being responsive to visitors' ideas and inquiries.
- Investigate more about how appropriate learning theory can support an orientation to science as inquiry.
- Continue to work on a theory of PCK for inquiry development.

Acknowledgments The authors thank Arthur Earle, Lisa Porter, Jill Sanderson, Christian Thomas, Bill Watson, and Nicole Webster for helping to design, launch, and sustain the program described in this chapter. We thank Jenna Carlson, Matty Lau, Jonathan Eakle, Lara Smetana, Kathleen Smith, and Binyu Yang for their contributions to this work. Finally, we thank the participating teacher candidates for contributing their journals and interview responses. This work was supported in part by a 100kin10 Collaboration Grant, as well as a grant from the National Science Foundation, Building Capacity for Disciplinary Experts in Math and Science Teaching, DUE 1439819. The views expressed are those of the authors and not necessarily shared by the Foundation.

References

Adams, J. D., & Gupta, P. (2017). Informal science institutions and learning to teach: An examination of identity, agency, and affordances. *Journal of Research in Science Teaching, 54*(1), 121–138.

American Association for the Advancement of Science (AAAS). (1993). *Benchmarks for science literacy*. New York: Oxford University Press.

Aquino, A. E., Kelly, A. M., & Bayne, G. U. (2010). Sharing our teachers: The required graduate class at the American Museum of Natural History for Lehman College (CUNY). *The New Educator, 6*(3–4), 225–246.

Avraamidou, L. (2014). Developing a reform-minded science teaching identity: The role of informal science environments. *Journal of Science Teacher Education, 25*(7), 823–843.

Center for Science, Mathematics, and Engineering Education (CSMEE). (2000). *Inquiry and the National Science Education Standards: A guide for teaching and learning*. Washington, DC: National Academy Press.

Chin, C. C. (2004). Museum experience—A resource for science teacher education. *International Journal of Science and Mathematics Education, 2*(1), 63–90.

Coffey, J. E., & Edwards, A. R. (2015). The role subject matter plays in prospective teachers' responsive teaching practices in elementary math and science. In A. Robertson, R. Scherr, & D. Hammer (Eds.), *Responsive teaching in science and mathematics*. New York: Taylor and Francis.

Duschl, R. (2008). Science education in three-part harmony: Balancing conceptual, epistemic, and social learning goals. *Review of Research in Education, 32*(1), 268–291.

Davis, E. A., & Krajcik, J. S. (2005). Designing educative curriculum materials to promote teacher learning. *Educational Researcher, 34*(3), 3–14.

Duckworth, E. (1996). *The having of wonderful ideas and other essays on teaching and learning*. New York: Teachers College Press.

Furtak, E. M., Seidel, T., Iverson, H., & Briggs, D. C. (2012). Experimental and quasi-experimental studies of inquiry-based science teaching: A meta-analysis. *Review of Educational Research, 82*(3), 300–329.

Gess-Newsome, J. (2015). A model of teacher professional knowledge and skill including PCK. In A. Berry, P. Friedrichsen, & J. Loughran (Eds.), *Re-examining pedagogical content knowledge in science education* (pp. 28–42). New York: Routledge.

Grooms, J., Enderle, P., & Sampson, V. (2015). Coordinating scientific argumentation and the next generation science standards through argument driven inquiry. *Science Educator, 24*(1), 45–50.

Hill, H., Ball, D., & Schilling, S. (2008). Unpacking pedagogical content knowledge: Conceptualizing and measuring teachers' topic specific knowledge of students. *Journal for Research in Mathematics Education, 39*(4), 372–400.

Jung, M. L., & Tonso, K. L. (2006). Elementary pre-service teachers learning to teach science in science museums and nature centers: A novel program's impact on science knowledge, science pedagogy, and confidence teaching. *Journal of Elementary Science Education, 18*(1), 15–31.

Kelly, J. (2000). Rethinking the elementary science methods course: A case for content, pedagogy, and informal science education. *International Journal of Science Education, 22*(7), 755–777.

Kuhn, D., & Pease, M. (2008). What needs to develop in the development of inquiry skills? *Cognition and Instruction, 26*(4), 512–559.

Levin, D., Hammer, D., & Elby, A. (2012). *Becoming a responsive science teacher: Focusing on student thinking in secondary science*. Arlington: National Science Teachers Association.

Mayer, R. E. (2011). *Applying the science of learning*. Boston, MA: Pearson/Allyn & Bacon.

Metz, K. E. (1995). Reassessment of developmental constraints on children's science instruction. *Review of Educational Research, 65*(2), 93–127.

Mortimer, E., & Scott, P. (2003). *Meaning making in secondary science classrooms*. Philadelphia: Open University Press.

NGSS Lead States. (2013). *Next generation science standards: For states, by states*. Washington, DC: The National Academies Press.

National Academies of Sciences, Engineering, and Medicine (NASEM). (2015). *Science teachers learning: Enhancing opportunities, creating supportive contexts*. Committee on Strengthening Science Education through a Teacher Learning Continuum. Board on Science Education and Teacher Advisory Council, Division of Behavioral and Social Science and Education. Washington, DC: The National Academies Press.

National Research Council (NRC). (2012). *A framework for K–12 science education: Practices, crosscutting concepts, and core ideas*. Washington, DC: The National Academies Press.

National Research Council (NRC). (2007). Taking science to school: Learning and teaching science in grades K-8. Committee on Science Learning, Kindergarten Through Eighth Grade. In R. A. Duschl, H. A. Schweingruber, & A. W. Shouse (Eds.), *Board on Science Education, Center for Education. Division of behavioral and social sciences and education*. Washington, DC: The National Academies Press.

National Research Council (NRC). (2000). *Inquiry and the National Science Education Standards: A Guide for Teaching and Learning*. Washington, DC: The National Academies Press. https://doi.org/10.17226/9596

National Research Council (NRC). (1996). *National science education standards: Observe, interact, change, learn*. Washington, DC: National Academy Press.

Rivera Maulucci, M. S., & Brotman, J. S. (2010). Teaching science in the city: Exploring linkages between teacher learning and student learning across formal and informal contexts. *The New Educator, 6*(3–4), 196–211.

Robertson, A. D., Scherr, R., & Hammer, D. (Eds.). (2015). *Responsive teaching in science and mathematics*. New York: Routledge.

Schwab, J. J. (1958). The teaching of science as inquiry. *Bulletin of the Atomic Scientists, 14*(9), 374–379.

Seligmann, T. (2014). Learning museum: A meeting place for pre-service teachers and museums. *Journal of Museum Education, 39*(1), 42–53.

Shulman, L. S. (1986). Those who understand: Knowledge growth in teaching. *Educational Researcher, 15*(2), 4–14.

Smetana, L., Birmingham, D., Rouleau, H., Carlson, J., & Phillips, S. (2017). Cultural institutions as partners in initial elementary science teacher preparation. *Innovations in Science Teacher Education, 2*(2). http://innovations.theaste.org/cultural-institutions-as-partners-in-initial-elementaryscience-teacher-preparation/

Stein, F., & Rankin, L. (1998). Developing a community of practice: Exploratorium Institute for Inquiry. *Journal of Museum Education, 23*(2), 19–21.

Sunderland, M. E., Klitz, K., & Yoshihara, K. (2012). Doing natural history. *Bioscience, 62*(9), 824–829.

Thompson, J., Windschitl, M., & Braaten, M. (2013). Developing a theory of ambitious early-career teacher practice. *American Educational Research Journal, 50*(3), 574–615.

Wood, E. (2003). Pedagogical content knowledge: An example from secondary school mathematics. *The Mathematics Educator, 7*(1), 49–61.

Chapter 11
Engineering STEM Teacher Learning: Using a Museum-Based Field Experience to Foster STEM Teachers' Pedagogical Content Knowledge for Engineering

Matty Lau and Satbir Multani

Abstract Empowering students to utilize science and mathematics to solve the world's problems is needed to prepare them for their future. To do that, science and mathematics teachers must know how to integrate engineering into their classrooms. In this chapter, we describe a novel museum-based field experience for novice science and mathematics teachers to develop their pedagogical content knowledge for teaching engineering and disciplinary practices. This field experience used the Design Lab exhibit at the New York Hall of Science as a way to introduce these novices to design-based and student-centered learning. Through the iterations of this experience, we found it critical to scaffold novice teachers' understanding of how to use disciplinary practices and engineering ideas to help students learn. What worked and future directions will be discussed.

Keywords PCK · Engineering · Museum-based teacher education · STEM teacher learning · NGSS · Common Core · Teacher residency · Design · Circuits · Instructional design

11.1 Introduction

With the Next Generation Science Standards, or NGSS, science teachers need to know how to help students "actively engage in scientific and engineering practices and apply crosscutting concepts to deepen their understanding of the core ideas in these fields" (National Research Council, 2012, p. 9). These three dimensions of science proficiency, as the authors of the NGSS have termed it, have elevated

M. Lau (✉)
Teacher Learning Consultancy, New York, NY, USA

S. Multani
New York Hall of Science, Corona, NY, USA

© Springer International Publishing AG, part of Springer Nature 2018
S. M. Uzzo et al. (eds.), *Pedagogical Content Knowledge in STEM*,
Advances in STEM Education, https://doi.org/10.1007/978-3-319-97475-0_11

engineering and engineering practices in science classroom learning. These are now considered critical components of science teaching. The rationale for this elevation is that:

> ...the insights gained and interests provoked from studying and engaging in the practices of science and engineering during their K-12 schooling should help students see how science and engineering are instrumental in addressing major challenges that confront society today, such as generating sufficient energy, preventing and treating diseases, maintaining supplies of clean water and food, and solving the problems of global environmental change. (National Research Council, 2012, p. 9)

Though engineering is not explicitly called out in the Common Core Mathematics Standards, understanding the application of mathematical ideas and practices are (National Governors Association Center for Best Practices, Council of Chief State School Officers, 2010). Engineering is one avenue with which students can apply mathematical thinking by appropriately choosing and using mathematical tools to make sense of and address major challenges that confront society today.

In order to teach in this way, science and mathematics teachers need the know-how to meld engineering and their disciplines effectively in classroom instruction. Teachers must have the pedagogical content knowledge, or PCK, to help them expand beyond science or mathematics to include defining and delimiting engineering problems, designing solutions, and optimizing designs. The core of high school teachers' identities, the primary focus of their training and professional work, has traditionally been centered around their subject area (McLaughlin & Talbert 1990). Even if they may have engineering experience, high school science or mathematics teachers may struggle with incorporating engineering into classroom learning to meet their professional responsibilities to science or mathematics learning.

At the New York Hall of Science (NYSCI), we designed a museum-based field experience (the Design Lab field experience) to help novice secondary science and mathematics teachers integrate engineering with science or mathematics learning. We used the Next Generation Science Standards' stance on engineering to guide the development of our field experience. While there are differences between engineering and design, we used them interchangeably in this field experience due to the nature of the teacher learning work. We will discuss why in the section on the foundations of this field experience.

Novice secondary science and mathematics teachers who participated in this program were enrolled in a teacher residency program focused specifically on their PCK development. Integrating engineering work with disciplinary practices, such as modeling, can help high school science and mathematics teachers learn how to effectively use engineering to support science and math learning in their classrooms. As a result, a cornerstone of this work was on engaging teaching residents in the teaching and learning process of engineering and disciplinary practices. In this chapter, we will describe the principles of this field experience, what we learned in the experience design process, and where we could go next with this work.

11.2 Why Science Museums?

Science museums (including science and technology centers) have long been in the business of helping the public understand science and engineering. They provide experiences aimed at helping visitors find reasons to care about science and engineering by moving beyond traditional textbook and classroom study of phenomena and scientific ideas (Friedman, 2010). Museum exhibits and experiences are purposefully designed to spark curiosity and allow visitors to have control over their own meaning making (Bevan et al. 2010; Falk & Dierking, 2013; National Research Council, 2009).

At NYSCI, an exhibit on the engineering and design process, the Design Lab, which opened in 2014, engages visitors in "solving personally motivating problems via a creative design process" that allows them to wrestle with the disciplinary content most relevant to their problems (Bennett & Monahan 2013). Design Lab's entry into engineering is ripe with possibilities for students to engage in the deep, multidimensional learning envisioned by the NGSS or Common Core. This, in turn, presents a place where teachers can strengthen their understanding of how to integrate engineering into science or mathematics learning.

Science museums frequently offer programming to serve STEM teacher learning needs (Phillips & Wever-Frerichs, 2007), from programs to train new science teachers (e.g., Morentin & Guisasola 2015) to professional development for seasoned teachers (e.g., Duran, Ballone-Duran, Haney, Beltyukova, 2009). Science museums have much potential for fostering teachers' PCK by helping them learn routines to support students' inquiry learning (e.g., Olson, Cox-Peterson, McComas, 2001), use concepts and processes to help students make sense of the world around them (e.g., Jeanpierre, Oberhauser, Freeman, 2005), understand how to engage a learning process that puts students' ideas and questions at the center of that process (e.g., Rivera Maulucci, Brotman, Fain, 2015 and Saxman, Gupta, Steinberg, 2010), and use objects and places to motivate student inquiry about natural phenomena (e.g., Pickering, Ague, Rath, Heiser, & Sirch, 2012). However, a challenge remains in helping teachers effectively bridge between their students' learning in the museum and in the classroom that maintains students' excitement in and personal ownership of their learning.

NYSCI has a history of helping to train new STEM teachers (see the CLUSTER program described in Saxman et al. 2010). A key feature has been integrating the teachers' museum learning with the teacher training course work to develop inquiry-oriented, reflective practitioners of science teaching. Expanding on this, NYSCI partnered with the City University of New York-Hunter College (Hunter) and New Visions for Public Schools (New Visions) in a National Science Foundation-funded project (NSF grant #1238157) to develop and implement a teacher resident program called the Mathematics and Science Teacher Education Residency (MASTER) program. This close partnership meant that the program objectives, design decisions, and program implementation responsibilities were shared across all three partners.

11.3   What Is the MASTER Residency Program?

The Mathematics and Science Teacher Education Residency (MASTER) program was a 2-year master's degree and certification program to train and certify secondary science and mathematics teachers. NYSCI, Hunter, and New Visions collaborated on the design and implementation of the MASTER program. The focus was on developing the pedagogical content knowledge these novice secondary mathematics and science teachers. After a review of the literature on developing STEM teachers PCK, the three program partners collaboratively developed the following program objectives:

- Build residents' conceptual models of mathematical and scientific phenomena to support their students in developing conceptual understandings.
- Residents will know key concepts and practices secondary students must learn in science/mathematics and articulate how students will make progress in their understanding over time.
- Residents will understand how students learn their content area and anticipate prior knowledge, initial thinking or misconceptions, and skills students bring to a task.
- Residents will employ multiple instructional models and approaches to help students access and understand disciplinary concepts and practices.
- Residents will make instructional choices based on student thinking and addressing conceptual or disciplinary skill gaps.
- Residents will use formative assessment to continually evaluate and monitor student understanding of disciplinary concepts and practices.

An intentional part of the program design was to blend the practical with the theoretical aspects of teacher learning throughout residents' training. In other words, we deliberately integrated the clinical and the coursework aspects of the residency so that the residents' work in one could inform the work in the other. The first year was organized around the residency work and the second year around the induction work. The Design Lab field experience was part of a suite of program work the residents completed in their first summer in the MASTER program to prepare for their residency work.

The Design Lab field experience incorporated NYSCI's Design-Make-Play pedagogical ideas (Honey & Kanter, 2013) into the MASTER residents' training: helping learners to use the design process to engineer innovative solutions to problems that matter to them, developing learners' maker identities through having them build or adapt things for the simple pleasure of figuring out how to make something work, and fostering learner's sense of playfulness as they explore or invent things.

11.4 What Is Design Lab?

In 2014, NYSCI opened Design Lab, a 6500-square foot permanent exhibition located in the lower central pavilion of the museum. This exhibition consists of five separate spaces or "pods" where visitors use engineering and design as they engage in activities and challenges. At the time, each pod was dedicated to different thematic activities:

- *Backstage* visitors create prototypes for performance-related challenges or problems (e.g., make jointed shadow puppets).
- *Sandbox* visitors construct "larger-scale" prototypes to tackle challenges often associated with structures and buildings (e.g., use wooden dowels and rubber bands to create a sturdy structure big enough for a person to stand inside).
- *Studio* visitors build tabletop prototypes to tackle challenges that may require attention to details such as what one would find in a design studio that models prototypes with scaled-down versions (e.g., build something out of cardstock and lit LEDs to contribute to a model city on display).
- *Treehouse* a split-level area where visitors tackle design challenges that require experimentation with vertical space (e.g., create a method to move objects between the two levels).
- *Makerspace* (opened in 2012) visitors use tools to make real, functional prototypes, rather than the mock-ups that are typically created in the other Design Lab pods (e.g., fabricate their design ideas or repurpose everyday materials in new ways).

With the exception of the Maker Space, the themes in each pod have changed over time. Currently, the pod activities have been restructured to encompass imaginative design, engineering, free play, and co-learning.

Design Lab's goal is to deeply engage all types of learners in solving personally motivating problems through the design process. By empowering the visitors to see design opportunities in the world around them, the visitors will be motivated to explore STEM topics in a way that is engaging to them. Design and engineering fits well within the mission of the museum; this mission is encapsulated in the phrase Design-Make-Play. NYSCI's Design-Make-Play approach aims to foster generations of passionate learners, critical thinkers, and active citizens (New York Hall of Science 2018, Mission Statement). Playing is the visitor's entry point into exploration and learning, making is honing in on engagement through skill building, and designing is using making and playing to inform, plan, and problem-solve.

11.4.1 Foundations of the Design Lab Field Experience

In the decades since Shulman launched the idea of pedagogical content knowledge (Shulman, 1986, 1987), much has been written about the specific professional knowledge that teachers use in teaching specific subject matter to their students and assessing that learning. Researchers have examined how this knowledge develops (e.g., Henze, van Driel, Verloop, 2008; van Driel, Verloop, de Vos, 1998), its nature, subfeatures, and types (e.g., Magnusson, Krajcik, Borko, 1999; McNeill, González-Howard, Katsh-Singer, Loper, 2015) as well as how to gather evidence of and measure it (e.g., Kirschner, Taylor, Rollnick, Borowski, 2015).

Additionally, research has shown some intervention features are effective at improving teachers' PCK. Activities that closely align with practice (e.g., field work, rich case studies, examining student work) tend to be more represented in studies that show effectiveness of interventions on teachers' PCK development (Daehler et al., 2015; Evens, Elen, Depaepe, 2015). Content knowledge development is necessary but not sufficient for PCK development (Daehler, Heller, Wong, 2015). In helping teachers develop their modeling PCK, activities needed to acknowledge where teachers are with respect to their knowledge and practice while at the same time provide perspectives that are distinct from what they are used to doing (Justi & van Driel, 2005). Engaging in higher-order thinking and reflecting on teaching and learning work, both individually and collectively, are shown to be effective in developing teachers' PCK and helping them move from knowing about to knowing how to enact effective teaching practice (Evens et al., 2015) or specifically engineering teaching practice (Sun & Strobel, 2014). Secondary teachers often use content learning to evaluate student learning of disciplinary practices and therefore need to develop their PCK of disciplinary practices with an eye on how disciplinary content understanding develops (McNeill & Knight, 2013). Teachers have an easier time connecting with the prototyping and redesign process than the other aspects of the engineering/design process, such as defining a problem (Hynes, 2012).

Two activities seemed critical for developing teachers' PCK: (1) experiencing and (2) reflecting. Based on research findings for these activities, we formulated design principles to guide our field experience development (see Fig. 11.1).

Design Lab Field Experience Design Principles
1. The field experience should immerse residents in a science learning experience that engages them in both the engineering and disciplinary modeling processes;
2. The field experience will capitalize on the strengths and resources the teachers bring (i.e., prototyping in the engineering process) to facilitate their learning;
3. The field experience will provide residents with opportunities and scaffolds to plan for teaching that supports student engagement in both the engineering and disciplinary modeling processes;
4. The field experience will provide residents with opportunities and scaffolds to enact teaching that supports student engagement in both the engineering and disciplinary modeling processes;
5. The field experience will provide residents with opportunities and scaffolds to engage in higher-order reflection on practice (e.g., individually and collectively analyze teaching and learning episodes);
6. The field experience will take advantage of the affordances of museums for learning (e.g., collaborative, personally meaningful learning).

Fig. 11.1 Design Lab field experience design principles

The MASTER program's Summer I learning goals provided additional guidance on formation of the field experience. Because this particular Design Lab field experience was part of a suite of MASTER program work to prepare the residents for their residency work in the fall, the Design Lab field experience needed to:

- Provide residents with a common experience of science learning that aligned with the Next Generation Science Standards (NGSS, 2013).
- Help residents understand and use STEM Education concepts, tools, and routines (e.g., project-based learning, 5Es, learning progressions, NGSS, theories on how people learn).
- Build the foundation for enacting the professional norm of evidence-based reflection on teaching.

As part of the MASTER program work, each resident received an iPad kit from NYSCI to use for the duration of their time in the MASTER program. Each kit included a microphone and preloaded applications to help residents video record their teaching work. The residents used these iPad kits during the Design Lab field experience to create evidence-based reflections on teaching and to practice creating video recordings for their New York State certification requirements (i.e., EdTPA).

Lastly, while we recognize the differences between the fields of design and engineering, they functionally aligned, in many ways, in this field experience. In the kind of design challenges and instructional work the teachers encountered and were expected to do, the residents needed to flexibly use the skills and mindsets from both fields. To that end, we used engineering and design processes interchangeably.

11.5   NYSCI Design Lab Field Experience

In this section, we present the design of the field experience, which underwent several iterations. The final iteration was a 32-h (or eight 4-h sessions) museum-based field experience that was divided into three major strands of learning. In the first strand of work, residents were immersed in a design challenge that embodied the ideas of Design-Make-Play. Pedagogically oriented reflection activities, where residents could learn to see pedagogical situations through a teacherly lens, formed the second strand of the teacher learning. In the final strand of learning, the residents engaged in an instructional design process to hone their skills as evidence-based instructional designers. While more work can be done, we believe the features of this design holds promise for building science teachers' PCK of engineering.

11.5.1   Design Challenge: Happy City

Design principles 1, 2, and 6 guided the design of this portion of the field experience (Fig. 11.2).

Design Principles for Residents' Design Challenge Experience

Design Principle 1: The field experience should immerse residents in a science learning experience that engages them in both the engineering and disciplinary modeling processes.

Design Principle 2: The field experience will capitalize on the strengths and resources the teachers bring (i.e., prototyping in the engineering process) to facilitate their learning.

Design Principle 6: The field experience will take advantage of the affordances of museums for learning (e.g., collaborative, personally meaningful learning).

Fig. 11.2   Design principles for residents' design challenge experience

Fig. 11.3   Residents working in Design Lab. (Photo credit: Todd Narasuwan)

The residents launched into an 8-hour design challenge experience with a "field trip" to the Studio Space in Design Lab to tackle the Happy City challenge (Bennett & Monahan, 2013). Residents were charged with creating something to increase happiness in Happy City. They were only allowed to use the materials available (i.e., cardboard, tape, pipe cleaners, light-emitting diodes (LEDs), 3 V watch batteries, and markers) to prototype their creations (see Fig. 11.3). On display in Studio Space were Happy City creations made by other visitors. Based on our experiences with visitors at the museum and our understanding of the research (i.e., Hynes, 2012), we decided that jumping into prototype and iterating was an easy and quick way to excite the residents' interest in the design challenge.

As part of the activity initiation, the facilitator (a museum staff specially trained for working with the MASTER residents) showed them the "Happy City Hot Tips" (see Fig. 11.4) and encouraged the residents to start by sketching initial ideas and to share what they figured out about the materials and how to get things to work. Additionally, the facilitator did not fix any problematic designs but directed residents to look at each other's work or to look at what was on display to find solutions to their problems.

> *Happy City Hot Tips*
> 1. *Make sure you have good metal-to-metal connection.*
> 2. *The long end of the LED needs to be connected to the positive side of the battery.*

Fig. 11.4 Happy City Hot Tips initial list

These Hot Tips were intentionally scant. While they provided residents with enough information to light an LED, this was not enough to help them troubleshoot all of the circuitry problems they might encounter. The facilitation and Hot Tips were designed to purposefully help residents frame their roles as collaborative sense-makers as they created working prototypes. In each iteration of the Design Lab field experience, we observed strong evidence of this collaborative sense-making when we saw residents troubleshoot each other's non-working designs and praise each other for their ingenuity in solving common problems.

A critical piece of the scientific modeling work in this experience involved activities wherein the residents communicate about their designs and what they understood from working with the materials. The Design Lab field experience instructor guided residents to (1) create blueprints of their working prototypes so future Happy City contractors would know how to build the designs and (2) add to the Hot Tips list to help the community of Happy City designers (e.g., future visitors to the exhibit). In order to effectively communicate their designs to these fictional contractors, the residents created a common visual language, or key, for representing critical features of their prototypes in their blueprints (e.g., what indicated tape versus aluminum foil/wiring). Creating this common key served an important purpose in the collaborative modeling process because the residents came to a consensus on what features were critical in representing the various working circuits in their prototypes.

Early on in our design work of the field experience, we determined that residents needed dedicated space and time to abstract their working knowledge of circuits. In the first iteration, residents were asked to add to the Hot Tips list as they worked on their designs. However, this did not happen. The residents were focused on getting functional designs rather than creating repositories of shared knowledge. Knowledge sharing occurred in informal ways, as they helped each other fix non-working designs or showed each other their creations. However, in the first iteration of the field experience, residents were reluctant to codify their burgeoning knowledge of circuits in the midst of creating their designs. This codification was a critical component of the communal sense-making of circuits. In subsequent Design Lab iterations, we set aside time for residents to add to the Hot Tips list so we could capture residents' understanding of the materials and generate a shared working model of circuits.

During this "adding to the Hot Tips" time, the residents reflected on what problems they encountered as they worked with the materials, what solutions they tried and categorized the various solutions into different types. Once categories of

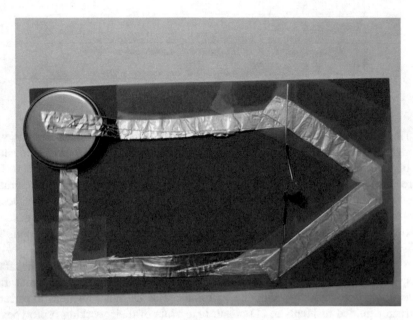

Fig. 11.5 Short-circuit example used in a formative assessment task

solution types were created, residents generated simple advice for future designers that would be added to the Hot Tips list (e.g., aluminum foil can extend the reach of the LED legs to the battery) and suggestions of what to avoid (e.g., avoid creating a path of aluminum foil back to the battery that does not have an LED in that path). The instructor highlighted these Hot Tips as part of their working model for how residents thought circuits with LEDs worked.

To assess how well residents could apply their understanding of circuits and troubleshooting, which is a key part of the design optimization process, residents were asked to "help Matty figure out why three of her four prototypes, which were supposed to follow the same design, did not work and figure out how to fix it." The residents were encouraged to closely examine, even take apart, the four designs (see Fig. 11.5). Three of the four circuits contained a short that prevented the LED from lighting. In the fourth, a piece of clear tape interrupted the path of the short. This formative assessment task was designed to see how well residents could flexibly apply their understanding to a novel situation and indicated whether residents needed to deepening their understanding of conductive paths and resistance or were ready to continue on to parallel circuits.

Individually, residents examined the circuits and wrote down their initial ideas. After, residents discussed their ideas in small groups or with partner. During this discussion time, they were encouraged to play with or even take apart the circuits to test out some of their ideas. Typically, residents could identify the short circuit problem in three of the four non-working circuits but could not see how the fourth circuit worked. The instructor scaffold, at this point, connected the discussion with their Hot Tip of

"Make sure you have good metal-to-metal connection." A piece of tape interrupted the metal-to-metal connection, which did away with the short circuit.

A final challenge was introduced that continued with the Happy City storyline. The MASTER residents learned that, while the citizens of Happy City liked the residents' creations, the citizens wanted the residents to redesign them with the intent of making the designs more environmentally responsible or "green." To solve this challenge, residents needed to clarify the problem, determine the design criteria, and identify the constraints.

The instructor guided the discussion so the residents would consider what efficient use of power might mean. All the iterations that led residents to explore parallel circuits and to notice that LEDs connected in series to a single battery did not work. For safety reasons, the residents were warned to not connect too many LEDs to the battery at the same time. LEDs in parallel do not mean lower power consumption. However, by building parallel LED circuits, the residents were able to see that they could light many LEDs, but the batteries died more quickly. This allowed the residents to compare and contrast the pros and cons of parallel and series LEDs as well as develop a conceptual model of parallel circuits. This modeling work resulted in additions to their Hot Tips list for working with these materials as well as surfaced questions they had about how circuits worked (e.g., why LEDs in series did not light up but incandescent bulbs, like strings of decorative lights, did).

At the completion of the "greenification" challenge (i.e., residents created new prototypes), the residents created schematics of their new designs and added to the Hot Tips list. Then the instructor discussed the physics she saw the residents developing in the course of working on the design challenges and where one could go in a typical high school physics class to continue students' learning of circuits (e.g., using incandescent light bulbs to develop a quantitative model of circuits and examining the power usage issue they encountered in the greenification challenge).

11.5.2 Residents Reflecting on the Design Lab Experience

Design principles 5 and 6 served as the main guides for this portion of the field experience (Fig. 11.6).

This 8-hour reflection work (spread over two sessions) served to bridge the residents' experience as STEM learners with their experience as STEM teacher learners. There were two main reflection activities, residents reflected on (1) their own design challenge experience and (2) middle school museum visitors engaging with the Design Lab activities. This helped to build the foundation for the residents' skills in higher-order reflection on teaching practice.

To foster evidence-based reflection, residents examined a low inference transcript (LIT) of their design challenge work where they diagnosed the non-working prototypes created for the experience. LITs are "are highly detailed transcripts of classroom... conversations, similar to a stenographic report" (Rivera-McCutchen & Scharff Panero, 2014, p. 87). From the LIT, the residents generated evidence of

> **Design Principles for Resident Reflection Learning**
>
> *Design Principle 5:* The field experience will provide residents with opportunities and scaffolds to engage in higher-order reflection on practice (e.g., individually and collectively analyze teaching and learning episodes).
>
> *Design Principle 6:* The field experience will take advantage of the affordances of museums for learning (e.g., collaborative, personally meaningful learning).

Fig. 11.6 Design principles for resident reflection learning

formative assessment strategies and residents' use of models and modeling to make sense of electric circuits. Residents also marshaled evidence for a final category of learning they cared about for their students that was beyond content mastery, such as collaborative problem-solving or perseverance. This final category was determined in collaboration with the residents in each iteration because it was meant to highlight a valuable aspect of learning they felt museums could promote. In each iteration, at least one resident publicly argued that high school students could not do this kind of work. This was always left as an open question that we would continue to explore.

Museums afford opportunities for teachers to see other important aspects of student engagement and learning, such as personally meaningful learning, collaborative problem solving, and persevering in the face of challenges during learning. The second reflection activity was organized around an opportunity to study visitors in the Design Lab exhibit to help the MASTER residents assess student learning beyond content mastery. Local middle school students were invited to NYSCI for a special visit to the Design Lab Exhibit so residents could observe those students in the exhibit space.

The residents created LITs of middle school students tackling the different design challenge activities in Design Lab. These LITs were scrutinized for evidence of visitors collaborating, independently problem solving, or taking ownership over their task. In the first iteration, the residents struggled with what counted as reasonable evidence of these things. At the residents' request, we drafted lists of criteria for evidence of collaboration, independently problem-solving, or student ownership of task, which we called our "look for" lists. These drafts were revised several times during the course of this reflection activity. Creating a consensus on what were our "look for" lists (or criteria for evidence of X) served to be a powerful tool for common understanding and for sharpening residents' skills in identifying appropriate evidence. As a result, this scaffold was incorporated into later iterations of the Design Lab field experience.

11.5.3 Residents as Instructional Designers

The design principles 3, 4, 5, and 6 guided the design of the final 16 hours of the Design Lab fieldwork (Fig. 11.7).

Design Principles for Instructional Designer Learning

Design Principle 3: The field experience will provide residents with opportunities and scaffolds to plan for teaching that supports student engagement in both the engineering and disciplinary modeling processes.

Design Principle 4: The field experience will provide residents with opportunities and scaffolds to enact teaching that supports student engagement in both the engineering and disciplinary modeling processes.

Design Principle 5: The field experience will provide residents with opportunities and scaffolds to engage in higher-order reflection on practice (e.g., individually and collectively analyze teaching and learning episodes).

Design Principle 6: The field experience will take advantage of the affordances of museums for learning (e.g., collaborative, personally meaningful learning).

Fig. 11.7 Design principles for instructional designer learning

The focus of the residents' learning was on a process for instructional design. Residents worked in content area teams, or mini-unit design teams, to improve upon mini-units (three to seven lessons around a major science or mathematics concept) they created. As part of this iterative process, residents led learning activities with museum visitors and used data from that enactment to refine their instructional plans.

The first stage of this work involved creating engineering challenges to motivate students' work and incorporating modeling practices to structure students' learning. Rather than having residents create new lesson plans, we instructed them to select from mini-unit ideas they created that summer for their methods courses. We hoped this would increase the chances of residents using what they learned at NYSCI in their residency work because these mini-units were designed to align with curricular objectives expected for the courses the residents would teach in the fall. Additionally, we felt it was more worthwhile for residents to learn how to adapt existing curricular resources rather than generate new ones because that is likely what they would do during the residency.

Several scaffolds were used to help residents with this initial redesign work. The residents used the "Generating Authentic Design Problems" protocol (see Fig. 11.8) to identify engineering challenges that had high potential for being interesting to students (Bennett & Monahan, 2013). Residents also explored the museum resources to spark inspiration for design challenges. Some design challenges residents generated were: helping conservationists set up the best nature preserve in a particular geographic region for ensuring maximum biodiversity and helping lighting engineers determine the optimal lighting configuration for a popular musician's upcoming music and dance performance.

The next stage of residents' instructional design work entailed cycles of implementing lesson activities, collecting data on enactment, and analyzing that data to

Generating Authentic Design Problems	
Given your core/major idea, brainstorm the characters and problems they might encounter in situations/settings.	
Settings (Places or situations that students might encounter or be interested in)	Characters (At least 3-6 characters who might be part of this setting)
Potential Problems (At least four problems to solve in this setting)	STEM Core/Major Ideas & Practices| ▼ (Ideas and practices need to wrestle with to solve the problems)

Fig. 11.8 Generating authentic design problems protocol

improve their mini-unit designs. While one of the mini-unit team members led summer camp visitors through one of the activities from their mini-unit, the other team members either videoed or created an LIT of the enactment. Using that data, residents analyzed how well their activity design worked and collaboratively decided on what to change. We went through as many cycles as necessary to ensure that all residents had a chance to implement, video, or produce an LIT.

The Data-Driven Noticing Protocol was a key scaffold during this stage of work (see Appendix). To create this scaffold, we modified the Data-Driven Dialogue Protocol (Teacher Development Group, 2002) to help residents focus on using what they noticed of student doing or thinking to inform redesign decisions. From our observations, the Data-Driven Noticing Protocol helped residents not rush to conclusions about enactment, resolve disagreements, and articulate their rationale behind proposed changes. The repeated cycles of enactment and redesign provided the residents with several opportunities to see how students (i.e., the summer camp visitors) engaged with the engineering challenges and model-eliciting activities. In the feedback from the end of each summer, the residents consistently found these repeated opportunities useful. Additionally, from anecdotal reports received from residents during their residency, at least two residents, from each year, reported using the tools from the Data-Driven Noticing Protocol to support their reflection on their own teaching.

11.6   Reflections

Museums can help science and mathematics teachers grow and strengthen their pedagogical content knowledge for integrating engineering into mathematics or science instruction. We found that scaffolds for helping novice teachers see how engineering and science practices fit together to support teaching and learning were necessary in our Design Lab field experience to realize this goal. The residents' feedback indicated that they appreciated being shown how this kind of teaching worked and having the opportunity to collaborate with others to try it out themselves. Additionally, from the different iterations of this field experience, the explicit scaffolds for the professional thinking and design work needed to integrate engineering and modeling practice improved what the residents' burgeoning teaching practice. While our data was limited in scope, this suggests a promising direction worth exploring. A next step is to design research studies to examine the efficacy these kinds of scaffolds for helping science and mathematics teachers integrate modeling and engineering in their own classroom teaching. Once we start looking into teachers' own classroom contexts, we can also begin investigating the contextual boons and barriers to teaching this way in K-12 classrooms. Such an exploration can help teacher educators identify what teachers may need to take this kind of work forward into their own classrooms.

Acknowledgments   The authors would like to thank Michaela Labriole, Shannon MacColl, and Carlos Romero for their thoughts and feedback. This material is based upon work supported by the National Science Foundation under grant number DRL-1238157.

Appendix: Data-Driven Noticing Protocol

Observations

In this phase, note only the facts observed in the data, not conjectures, explanations, and conclusions. Make statements about quantities (e.g., over half the students, etc.) and the presence of specific information (e.g., Jabeer said "the earth shook because the magma boiled").

Individually record: what students said or did in relation to the focus of today's redesign work (e.g., collaboration, learning objectives, modeling practice)? If you catch yourself using "...because...," "it seems...," or other inference statements, then stop. Here are some helpful thought starters:

- I observe that...
- This student said/did (or didn't say/do)...

Discuss with group members (1) similarities and differences in your observations. Please be sure to help each other stay focused on observations and not inferences.

Inferences

In this phase, you will (a) interpret the observations, (b) generate multiple explanations for your observations, and (c) identify additional data that may be needed to confirm or contradict your explanations.

Please study the observations you made and privately record (1) what you learned about your students' {e.g., doing modeling} and what evidence was used to infer their {e.g., modeling work}. Here are some helpful sentence starters:

- I believe the data suggests… because…
- Additional data to help verify/confirm my interpretation/explanation would be…

Discuss with group members (1) similarities and differences in your inferences. Stay focused on using evidence from the observations. Be ready to ask clarifying questions and/or disagree if inferences do not seem supported by the data.

Implication

During the implications phase, you will make decisions about your plans based on your analysis.

Privately identify and record (1) (instructional redesign) what will/will not need to be modified and why and (2) (instructional next steps) if you were teaching these students tomorrow, what should your class do next and why?

Discuss with your group mates your instructional redesign and next steps proposals. Be sure to include a rationale for those suggestions and how that rationale is connected to your observations and inferences.

Observations	Inferences

Implications	Rationale
What to keep?	Based on your observations and inferences, why?
What to change?	Based on your observations and inferences, why?
What to do in future lessons in this unit/topic?	Based on your observations and inferences, why?

References

Bennett, D. & Monahan, P. (2013). NYSCI Design Lab: no bored kids! In M. Honey & D. E. Kanter (Eds.), *Design make play: Growing the next generation of STEM innovators* (pp. 34–49). New York, NY: Routledge.

Bevan, B., Dillon, J., Hein, G. E., Macdonald, M., Michalchik, V., Miller, D., Root, D., Rudder, L., Xanthoudaki, M., & Yoon, S. (2010). *Making science matter: Collaborations between informal science education organizations and schools.* Washington, DC: Center for Advancement of Informal Science Education.

Daehler, K. R., Heller, J. I., & Wong, N. (2015). Supporting growth of pedagogical content knowledge in science. In A. Berry, P. Friedrichsen, & J. Loughran (Eds.), *Re-examining pedagogical content knowledge in science education* (pp. 45–59). New York, NY: Routledge.

Duran, E., Ballone-Duran, L., Haney, J., & Beltyukova, S. (2009). The impact of a professional development program integrating informal science education on early childhood teachers' self-efficacy and beliefs about inquiry-based science teaching. *Journal of Elementary Science Education, 21*(4), 53–70.

Evens, M., Elen, J., & Depaepe, F. (2015). Developing pedagogical content knowledge: Lessons learned from intervention studies. *Education Research International, 2015*, 790417., 23 pages. https://doi.org/10.1155/2015/790417.

Falk, J. H., & Dierking, L. D. (2013). *Museum experience revisited.* Left Coast Press.

Friedman, A. (2010). The evolution of the science museum. *Physics Today, 63*(10), 45–51.

Henze, I., van Driel, J. H., & Verloop, N. (2008). Development of experienced science teachers' pedagogical content knowledge of models of the solar system and the universe. *International Journal of Science Education, 30*(10), 1321–1342.

Honey, M., & Kanter, D. E. (Eds.). (2013). *Design, make, play: Growing the next generation of STEM innovators.* New York, NY: Routledge.

Hynes, M. M. (2012). Middle-school teachers' understanding and teaching of the engineering design process: A look at subject matter and pedagogical content knowledge. *International Journal of Technology and Design Education, 22*(3), 345–360.

Jeanpierre, B., Oberhauser, K., & Freeman, C. (2005). Characteristics of professional development that effect change in secondary science teachers' classroom practices. *Journal of Research in Science Teaching, 42*(6), 668–690.

Justi, R., & van Driel, J. (2005). The development of science teachers' knowledge on models and modelling: Promoting, characterizing, and understanding the process. *International Journal of Science Education, 27*(5), 549–573.

Kirschner, S., Taylor, J., Rollnick, M., Borowski, A., & Mavhunga, E. (2015). Gathering evidence for the validity of PCK measures. In A. Berry, P. Friedrichsen, & J. Loughran (Eds.), *Re-examining pedagogical content knowledge in science education* (pp. 229–241). New York, NY: Routledge.

Magnusson, S., Krajcik, J., & Borko, H. (1999). Nature, sources, and development of pedagogical content knowledge for science teaching. In J. Gess-Newsome & N. G. Lederman (Eds.), *Examining pedagogical content knowledge* (pp. 95–132). Dordrecht, Netherlands: Kluwer Academic Publishers.

McLaughlin, M. W., & Talbert, J. E. (1990). The contexts in question: the secondary school workplace. In M. McLaughlin, J. Talbert, & N. Bascia (Eds.), *The contexts of teaching in secondary schools: teachers' realities* (pp. 1–14). New York, NY: Teachers College Press.

McNeill, K. L., González-Howard, M., Katsh-Singer, R., & Loper, S. (2015). Pedagogical content knowledge of argumentation: Using classroom contexts to assess high-quality PCK rather than pseudoargumentation. *Journal of Research in Science Teaching, 53*(2), 261–290.

McNeill, K. L., & Knight, A. M. (2013). Teachers' pedagogical content knowledge of scientific argumentation: The impact of professional development on K–12 teachers. *Science Education, 97*(6), 936–972.

Morentin, M., & Guisasola, J. (2015). The role of science museum field trips in the primary teacher preparation. *International Journal of Science and Mathematics Education, 13*(5), 965–990.

National Governors Association Center for Best Practices, & Council of Chief State School Officers. (2010). *Common core state standards mathematics.* Washington, DC: National Governors Association Center for Best Practices, Council of Chief State School Officers.

National Research Council. (2012). *A framework for K-12 science education: Practices, crosscutting concepts, and core ideas.* Washington, DC: The National Academies Press.

National Research Council. (2009). *Learning science in informal environments: People, places, and pursuits.* Washington, DC: The National Academies Press.

NGSS Lead States. (2013). *Next generation science standards: For states, by states.* Washington, DC: The National Academies Press.

New York Hall of Science. (2018). *Institutional mission statement.* Retrieved from https://nysci.org/home/about/

Olson, J. K., Cox-Petersen, A. M., & McComas, W. F. (2001). The inclusion of informal environments in science teacher preparation. *Journal of Science Teacher Education, 12*(3), 155–173.

Phillips, M., Finkelstein, D., & Wever-Frerichs, S. (2007). School site to museum floor: How informal science institutions work with schools. *International Journal of Science Education, 29*(12), 1489–1507.

Pickering, J., Ague, J. J., Rath, K. A., Heiser, D. M., & Sirch, J. N. (2012). Museum-based teacher professional development: Peabody fellows in earth science. *Journal of Geoscience Education, 60*(4), 337–349.

Rivera Maulucci, M. S. R., Brotman, J. S., & Fain, S. S. (2015). Fostering structurally transformative teacher agency through science professional development. *Journal of Research in Science Teaching, 52*(4), 545–559.

Rivera-McCutchen, R. L., & Scharff Panero, N. (2014). Low-inference transcripts in peer coaching: A promising tool for school improvement. *International Journal of Mentoring and Coaching in Education, 3*(1), 86–101.

Saxman, L. J., Gupta, P., & Steinberg, R. N. (2010). CLUSTER: University-science center partnership for science teacher preparation. *The New Educator, 6*(3–4), 280–296.

Shulman, L. S. (1986). Those who understand: Knowledge growth in teaching. *Educational Researcher, 15*, 4–14.

Shulman, L. S. (1987). Knowledge and teaching: Foundations of the new reform. *Harvard Educational Review, 57*, 1–22.

Sun, Y., & Strobel, J. (2014). From knowing-about to knowing-to: Development of engineering pedagogical content knowledge by elementary teachers through perceived learning and implementing difficulties. *American Journal of Engineering Education, 5*(1), 41–60.

Teacher Development Group. (2002). *Data Drive Dialogue.* Retrieved from: https://www.nsrfharmony.org/system/files/protocols/data_driven_dialogue_0.pdf

van Driel, J. H., Verloop, N., & de Vos, W. (1998). Developing science teachers' pedagogical content knowledge. Journal of Research in Science Teaching, 35(6), 673–695.

Chapter 12
Collaborative PCK in Practice: Bringing Together Secondary, Tertiary, and Informal Learning in a STEM Residency Program

Stephen Miles Uzzo, Harouna Ba, and Laycca Umer

Abstract In this chapter, the authors discuss the potential, value, and barriers to supporting PCK in the science, technology, engineering, and math (STEM) disciplines through collaborations across learning settings, with special attention to the role of informal science learning institutions in preservice teacher preparation. Particular focus is provided on the National Science Foundation (NSF)-funded Math and Science Teacher Education Residency (MASTER) program for teacher preparation. MASTER is a collaborative among the New York Hall of Science, Hunter College School of Education, and the New York City Department of Education through New Visions for Public Schools.

Keywords PCK · STEM · Inquiry-based learning · Design-based learning · Teacher education · Education residency · Informal science learning · Museum learning

12.1 Overview

The purpose of this chapter is to explore the value and role of informal learning institutions in preservice teacher clinical residency programs based on pedagogical content knowledge (PCK), in particular the partnership among New York City Department of Education schools (through New Visions for Public Education), the City University of New York (primarily the Graduate School of Education for Hunter College), and the New York Hall of Science (a city-owned, but privately operated public science museum—a hands-on laboratory and resource for teachers' professional growth). Emphasis will be placed on explicating the purpose of such collaboration, the structure and processes and challenges that arise in the

S. M. Uzzo (✉) · H. Ba · L. Umer
New York Hall of Science, Corona, NY, USA

complexity of collaborating in the execution of residency programs, and diagnoses of and adaptations around such challenges to best serve the diverse teachers and students in the country's largest school district. Some background in PCK and its value in preservice learning are provided and the role of the New York Hall of Science (NYSCI) in supporting PCK in teacher residencies for New York City (NYC) schools through inquiry-based learning and practice.

12.2 What Does Pedagogical Content Knowledge Mean in Practice?

Pedagogical content knowledge (PCK) is a sophisticated construct, requiring that the teacher deeply engage with the domain content, be flexible in teaching, understand multiple pathways into learning, and understand that learning is contextual to the setting and dependent on the habits of mind and prior knowledge of the learner (van Driel and Berry 2012). Effective STEM teachers who integrate PCK into their instruction are prepared to notice and assess student understanding and misconceptions, be responsive with individual learning while it is happening in the learning setting, and assist all students to increase knowledge and skills through an iterative process of articulation, making mistakes and reflecting, processing new ideas, engaging with questioning in collaboration with other students, and deepening discussion (Ball and Cohen 1999; Grossman et al. 2005; Remillard and Geist 2002). Ironically, particularly in high-needs schools, where motivation and achievement are low and class sizes are large, teacher feedback is typically negative, with little attention to individual learning and reinforcing the students' perceived inability to learn (Black et al. 2004).

The Mathematics and Science Teacher Education Residency (MASTER, NSF award number 1238157) program is intended to prepare teacher residency participants to use inquiry-based methods to encourage student discussion within the framework of responsive instruction, remove the stigma of failure, and replace it with students' exploration of their own and their classmates' thinking on conceptual questions. MASTER partners work closely with teachers in a 2-year residency-induction process to scaffold this transformation toward ambitious math and science instruction informed by PCK. This shift requires the integration of PCK into daily STEM instructional strategies (Ball et al. 2008a). Moreover, particularly in urban schools serving high-needs populations, teachers need the skills and knowledge to engage diverse student populations in science and math, which requires all teachers to understand and enact different approaches to accessing and understanding content. To be an effective teacher, development and preparation model for teachers in NYC schools, MASTER courses and ongoing coaching, and induction and professional development (PD) must integrate the building of PCK as an integral component (Moje et al. 2001; Lee 2004).

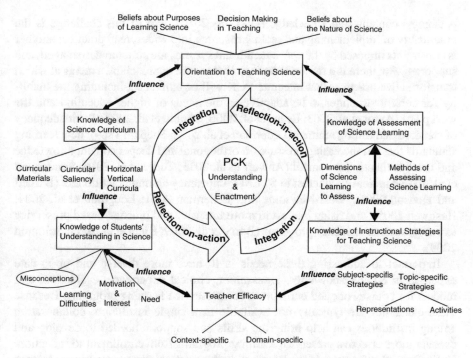

Fig. 12.1 Hexagon model of pedagogical content knowledge for science teaching. (Park and Oliver 2008a)

Since Shulman's watershed work articulated the need for a synthesis of content and pedagogical knowledge (Shulman 1986), there has been a great deal of research into what it means to combine knowledge and pedagogy. His focus on a systemic model for pedagogical reasoning and action illustrates the process of teaching practice as a feedback loop (see Fig. 12.1). Central to the notion of pedagogical content knowledge is integration: how we bring together teaching, learning, and content into a systemic whole (Wilkes 1994; Starkey 2010; Pella 2015). He was certainly not the first to advocate for a more integrated approach (Dewey 1902), but to firmly establish its importance as a way of thinking about the process of education as a system, rather than a set of separate frames.

12.3   Supporting PCK in Collaborative Environments

Two decades of research, hence, indicate the transformative potential of this model (Park and Oliver 2008b; Leach et al. 2006; Millar and Hames 2006; Loucks-Horsley et al. 2009), yet conventional teacher preparation, the means of assessing teacher performance, and student knowledge continue to embrace disparate models of pedagogy, content, and knowledge (Dikilitaş 2016). Part of this challenge is the practicality of implementing and scaling research into widespread practice; another is constraints imposed by the way educational systems are structured, managed, and supported. But there is a reason to believe that we are approaching a nexus at which transformative practices that integrate PCK will be embraced, including the inability for current structures to be adaptive to the needs of higher education and the workplace (Autor et al. 2003; Evans et al. 2006; Chubin et al. 2005); the inadequacy of standardized testing regimes (Lederman et al. 2002; Shepard 1989); new learning standards that put increasing pressure on orthogonal and deeper content knowledge and more authentic assessment (Anwari et al. 2015; Tuttle et al. 2016; Sondergeld et al. 2016); increasing interest in STEM competency among practitioners (Barnett and Hodson 2001a; Gess-Newsome and Lederman 2001; Loughran et al. 2004; Hashweh 2005); and rising interest in more intensive and practice-based preservice learning, such as residency programs (Berry et al. 2009; Klein et al. 2013; Solomon 2009).

Important to addressing these needs is to have more diverse and up-to-date expertise in STEM knowledge, assessment, clinical experience, evaluation, and models for coursework and curriculum for preservice learning programs. Because these capacities are typically not available in a single institution, collaboration among institutions can help bring the skills and supports needed to develop and execute more effective preservice learning programs from enrollment to induction. Institutions of informal science learning, such as natural history museums, science museums, and nature centers are useful in this regard (Chin 2004; Jung and Tonso 2006; Thirunarayanan 1997). In particular, informal institutions provide settings that allow for flexibility in the structure and process of learning (Doppelt and Schunn 2008), and provide opportunities to explore and validate new practices. And evidence suggests that informal environments can deepen and provide richer dimensions to both learning and teaching (Griffin 1994; Gerber et al. 2001; Cross 2007). For instance, inquiry learning techniques, such as project-based learning (PBL) and design-based learning (DBL), are used routinely in informal settings and are of particular value in deepening engagement in STEM learning both directly with students and in developing professional skills and STEM knowledge (Bransford et al. 1999; Brown et al. 1989; Edelson 2001; Geier et al. 2008; Mehalik et al. 2008; Fortus et al. 2004; Pacific Policy Research Center 2010). Leveraging the capacity for deeper learning available in informal learning environments for preservice teacher preparation may be able to aid the development of models for classroom organization and teaching methodologies. But aligning multiple institutions with divergent missions and business practices possesses challenges too. It is useful for the education research field to look at both the potential of these institutions, their role in innovative collaboratives and what we can learn from their implementation, both incipient and in the longer trajectory.

12.4 Collaboration in Residency Programs

While the MASTER program uniquely focuses on PCK, the New York Hall of Science has been collaborating in teaching residency programs with New York City Department of Education schools and the City University of New York for over a decade. Lessons learned through each successive program have informed the design of the next, and ultimately, led to the particular design of the MASTER program.

12.4.1 Cluster

Among the earliest of these programs was the Collaboration for Leadership in Urban Science Teaching, Evaluation, and Research (CLUSTER, NSF award number 0554269), in which the New York Hall of Science partnered with the City College of New York and the City University of New York Center for Advanced Study in Education. It engaged undergraduate students as inquiry-based interpreters (explainers) for visitors to the New York Hall of Science. Integrating this experience with a formal teacher certification program enabled participants to coordinate experiences in the science center, college science and education classes, and K-12 classrooms. The approach has its theoretical underpinnings in the concept of situated learning as noted by Kirshner and Whitson (Kirshner and Whitson 1997). Through apprenticeship experiences, situated learning recreates the complexity and ambiguity of situations that learners will face in the real world. Informal science institutions provide a potentially beneficial setting for situated learning by future teachers, allowing them to develop, exercise, and refine their science teaching and learning skills as noted by Gardner (Gardner 1991). Research indicated that CLUSTER graduates teaching in science classrooms outperformed the more experienced control group teachers (Saxman et al. 2010).

12.4.2 SPIRITAS

A more recent project is Science Pedagogy, Inquiry, and Research in Teaching Across Settings (SPIRITAS)—a collaborative among NYSCI, Queens College, and five New York City Department of Education high schools in the borough of Queens. SPIRITAS is a clinically rich program that was designed for individuals with an undergraduate degree in science who wish to enter the teaching profession. Pedagogy, science, and research in teaching are intertwined throughout the curriculum with a goal of graduating secondary science teachers with the knowledge skills and dispositions to (a) reflect on their practice, (b) integrate technology into their classroom, (c) address the learning needs of all students, and (d) work with students from diverse backgrounds and cultures. Candidates get their clinically rich experience early, introducing experiential and inquiry-based learning during their first summer (when they begin the program) and first fall with NYSCI. Fall and spring semesters of the first year also involve both coursework and practice, to be completed in participating schools. In the second year, the emphasis is on teaching. Candidates have their own classes to teach, and they also observe mentor teachers and attend professional development seminars. While results from this program are not yet available, it shows promise in terms of providing a deeper, richer, and more diverse preservice learning experience for educators as well as being generalizable to the clinic.

12.5 MASTER

Most recently, and with a specific focus on PCK, is the Math and Science Teacher Education Residency (MASTER) program, a collaborative effort among the Hunter College School of Education, the New York City Department of Education, New Visions for Public Schools, and the New York Hall of Science. It consists of a 2-year (residency year and induction year) Alternative Route to Certification program for New York City schools. In the context of MASTER, NYSCI aims to contribute science content and science pedagogy expertise to oversee the development and delivery of instructional experiences designed to build preservice residents' pedagogical content knowledge. The goals of NYSCI's participation in this project are twofold:

1. Provide a hands-on museum setting where residents can gain experience in (a) inquiry-based learning, which focuses on the concerns of teachers and is grounded in the realities of educational practice (Dana and Yendol-Hoppey 2009), and (b) design-based learning, which is an inquiry-based learning approach that uses design thinking and process in instruction (Kolodner et al. 1998).
2. Become a recognized and regularly used site for preservice teacher preparation by building the museum's capacity to work more closely and regularly with science and math faculty content experts from Hunter College and New Visions.

NYSCI's role in the MASTER program focused on supporting PCK development and scaffolding learning through the use of integrated, inquiry-based methods in the following areas:

1. *Field experience at NYSCI:* Development of residents' PCK in mathematics and science took place through integrated coursework and theoretical learning in constructionist activities and challenges in the museum workshop and gallery setting. Residents aligned curricula and units of instruction to this experience for school-year instruction. Resident feedback was used to collect evidence about the degree to which activities supported clinical PCK development for both content and disciplinary practices. In addition, specific issues and pinch points were documented and remediated throughout the experiences.

2. *Development of mentors and coaches:* In conjunction with field experience, NYSCI supported the development of mentors and coaches to help them scaffold PCK, both in the Hunter coursework and the clinical experience in New Visions schools. NYSCI also supported the residents' Defense of Learning (DOL) projects, in which residents present demonstrations of student learning using multiple measures and which function as summative performance assessments for all residents. NYSCI's role was to work with New Visions Director of Residencies in developing PCK "look fors," which provide evidence that residents are implementing PCK in their practice.
3. *Assistance with practice-focused coursework:* NYSCI also supported development of PCK in instructor's coursework at Hunter. This included developing new syllabi in collaboration with Hunter instructors, training through co-teaching with a focus on PCK, and development of assessment of resident performance using modifications to the Danielson Rubric, which identifies those aspects of teaching that have been proven through empirical studies and theoretical research as promoting improved student learning (Danielson 2013).

12.5.1   Field Experience

Residencies began with workshops in the museum setting in the first summer along with coursework. Museum resources and educators informed the design of the experiences. Residents were confronted with the pedagogical challenges of creating and enacting learning with the qualities of informal learning environments: enjoyable learning experiences, a breadth of levels of cognition, skills, and prior knowledge. Residents were compelled to take these qualities into account in their design of lessons and curricula with the intent of meeting specific learning outcomes.

Initially, residents participated in a summer-long optics exploration wherein they engaged in scientific argumentation to investigate and model the phenomenon of light on the museum floor. Through facilitation, they shared a lot of conflicting ideas, which led to deep discussions about what it means to know, how to elicit student thinking and possible misconceptions, and how to use that knowledge in the design of learning experiences; for example, using mirrors to explore optics by eliciting and testing hypotheses, e.g., that mirrors are like windows, so you would need a mirror the size of the thing you are trying to view, which turns out to be false (National Center for Teacher Residencies 2016).

Residents also worked in professional practice for developing and piloting an inquiry-based unit of instruction (3–5 days of lessons) and learning progressions to focus on core concept learning and to evaluate student interactions in the unit's learning activities. They further gathered data to support evaluation and revision of these units and identified the Next Generation Science Standards (NGSS) science practices most relevant to their lesson development, used a protocol to collect evidence to support decision-making about revisions, and revised them collaboratively within the resident cohort.

Working in design teams, residents developed engaging and accessible science activities for the museum floor and tested them with summer camp students possessing varying levels of science knowledge. The resident teams worked with mentors to plan activities involving a focus on student thinking and misconceptions including a substantive *check for understanding* component in a common lesson planning template. Residents were also expected to chronicle common misconceptions about the lesson topic using such things as aligning misconceptions with multiple-choice questions, recording misconceptions for each answer choice and engaging in error analysis with warm-up or exit tickets as part of their planning for future lessons [ibid]. Sessions were recorded for group review and critique for formative instructional improvement.

For year 3 of the project, it was determined that addition of a second summer experience for each cohort would be beneficial between the residency and induction years of the program. This second, hands-on field experience was designed and implemented as a follow-up to their original summer workshops to help reinforce and clarify disciplinary modeling in practice in support of resident learning of disciplinary content. This summer workshop focused on helping residents understand what disciplinary modeling (Rinke et al. 2014) in a math/science classroom looks like and what a teacher needs to do to support that kind of intellectual work. It included residents developing a set of engineering, design-based activities using basic materials that they created with the help of a NYSCI camp group. The workshop focused on creating effective problem frames and identifying questioning strategies that promote inquiry. Residents engaged with hands-on activities, evaluation of student work, and museum floor exploration.

Based on feedback, the summer workshop and training was redesigned for better integration of NYSCI summer experience and PCK development with residents' clinical experience. This aided greater alignment among different MASTER project elements. A Hunter College practicum field supervisor and NYSCI coaching staff observed each resident four times as part of conducting field coaching supervision. This integration of support increased alignment among the coursework, the clinic, and the field. Each resident was observed and given feedback two times per month by a coach.

In collaboration with New Visions, NYSCI developed and expanded the MASTER summer resident field experience as a credit-bearing, in-service, professional development course for New York City teachers. The scheduling and format were modified to accommodate teachers already in practice and made enhanced use of NYSCI's Design-Make-Play approach to STEM learning, resources, and professional development staff. Entitled "Design-Make-Play and Classroom Instruction: Creating Lesson and Assessment Plans," the professional development course is a blended learning workshop (a mix of in-person and asynchronous online work (Graham 2005)) in which participants discover how to develop instruction using NYSCI's *Design-Make-Play*, a STEM constructionist learning approach (Papert and Harel 1991). *Design-Make-Play* takes an instructional approach to design-based learning emphasizing personally meaningful real-life problems and engineering divergent solutions as a way to motivate learning (Honey and Kanter 2013).

In developing the course, NYSCI used *Design Lab*, an 8000-square-foot design and engineering exhibition at NYSCI, as the pathway into using constructionist ideas to support classroom science and math learning. Residents explored specific design challenges using readily available materials. For instance, an electrical circuit challenge used mostly familiar materials (cardboard, tape, light-emitting diodes [LEDs], aluminum foil, and watch batteries) to create something that will add to the "happiness" of a model city. Residents had to develop their own design, but in all cases it involved creating an electrical circuit to light LEDs, make sounds, and/or elicit movement. In working with the materials, they learned how to plan open-ended, inquiry-oriented lessons on learning how simple circuits work and how to construct stable physical structures. The pedagogy focused on (1) development of open-ended questions and framing problems in ways that engage students in the design process, (2) usage of the design process to support content instruction, and (3) assessment of student cognitive engagement and learning. Residents in this course drew on the NYSCI's rich curricular materials to create design-based lessons and assessments. The course was run during the spring semester, enrollment was high, and the course resources were made available to replicate the course in perpetuity.

During the break between fall and spring semesters (technically the Hunter College Graduate School of Education Winter Session), NYSCI and Hunter College collaboratively hosted science and mathematics residents from participating schools on an evening and weekend schedule for a clinical learning experience led by staff from NYSCI, New Visions, and Hunter. Working in disciplinary groups (living environment/biology, earth science, chemistry, and mathematics), they explored core concepts and PCK, gained practical experience with inquiry-based instruction, learned about integrating language development and literacy within STEM instruction and their curricula, and strengthened their understanding of mentoring and school improvement as they developed, taught, and revised their inquiry modules. Activities included:

- Conducting inquiry labs on core STEM concepts, consistent with National Research Council (NRC) and National Council of Teachers of Mathematics (NCTM) recommendations and Next Generation Science and Common Core State Standards.
- Customization and testing (in situ, in the clinic) of an initial inquiry-based module on a core concept in STEM learning.
- Training in supporting and assessing language and literacy development for adolescent learners.
- Training in coaching and collaboration techniques, including videotaping lessons, tagging and commenting on videos, and observation and structured reflection.
- Learning to use a portfolio evaluation system for residents and analyzing formative assessments of student learning for the inquiry modules.

Subsequently, a monthly 2-h spring inquiry seminar was implemented for residents by content area to develop their capacity in using assessment data to drive instruction. They met, reflected, and shared instructional strategies and provided peer feedback in content-specific groups to deepen their PCK around specific student learning needs and challenges within their content areas.

12.5.2 Development of Mentors and Coaches

Science coaches were recruited through the New Visions and Hunter College partnership with NYSCI and were a key form of support in the MASTER model for both residents and mentor teachers. In meetings with residents, NYSCI coaches used the DMP approach and a cognitive coaching model (Costa and Garmston 1994), which facilitated moving basic questions that residents brought to in-depth, data-driven discussions about pedagogy and student learning. Coaches led training in a systematic and disciplined process for interpreting and monitoring data and drawing actionable conclusions to apply in instructional practices. NYSCI coaches also supported mentor teacher' use of formative assessments (Honey and Kanter 2013) to monitor student learning, provided instructional feedback, and guided teachers in sharpening their own observation and mentoring skills.

NYSCI also designed and implemented science mentor teacher PCK professional development activities to create a learning community for mentors based on issues of science teaching or teacher mentoring. Coaches supported collaborative teams of science and math mentors through group problem-solving and reflection activities, including the use of the inquiry process for a mentoring challenge and receiving peer input and feedback. PCK-focused activities included:

1. Engaging in collaborative analysis of student work to uncover misconceptions and discuss instructional responses.
2. Learning new teaching strategies or instructional activities, discussing how and when to apply them, and how to support a resident in implementing them.
3. Norming around the definition and integration of Common Core math and science practices (Black and Wiliam 2009; National Governors Association Center for Best Practices 2010).
4. Discussing subject-specific indicators of more general best practices as described in the Danielson framework for teaching (Danielson 2013).

In addition, as needed, coaches facilitated three-way meetings between the resident, the mentor, and the coach. This occurred if a resident/mentor pair experienced challenges in their relationship. MASTER staff regularly communicated with mentors to discuss and monitor resident progress, address challenges that arose, and provide support with the development of mentor coaching skills.

Each science mentor was given the opportunity to present to the science mentor group the activities that they were working on with their students in science (e.g., helping students write better lab reports and develop their science literacy skills) or with their resident (e.g., helping the resident use diagrams to support conceptual learning in Earth Science) and discussed their work and solicited feedback. Coaches also joined as participants in these professional learning communities by presenting challenges encountered in their science teacher coaching work (e.g., in helping residents better incorporate modeling practice into their lesson planning work).

Mentor Teachers PCK Development All mentors participated in 20 h of professional development. For the second year, NYSCI designed face-to-face and online professional development (PD) activities to support mentors' PCK development in science teaching (e.g., more attention to student conceptual development and expanded inquiry-based teaching). The blended format was intended to provide more individualized feedback to mentor candidates, self-paced independent and partner work time, and greater opportunities to assess mentor competency prior to the beginning of the residency. Mentors engaged in 10 h of online individual and partner work to prepare for and practice skills developed during 10 h of in-person workshops. The face-to-face PD focused on developing PCK in areas of scientific practices (from NGSS) with mentors sharing their own teaching projects. Mentors used video analysis of classroom practices, research projects, and student work. The online PD was created and facilitated to focus on issues that mentors wanted to spend more time discussing with colleagues and to overcome mentors' reluctance to participate in professional development.

But feedback indicated that mentors did not expect to have to commit time for their own professional development; struggled with the complexity of PCK and activities such as the use of video reflection, identifying evidence of learning, and practice of scientific argumentation; found PCK generally theoretical and abstract; and had trouble translating it to practical mentorship in and about the clinic. Better communication of expectations for the mentor role, the effective use of blended learning, recasting PCK more through a mentorship lens, and helping mentors develop strategies to connect and contrast PCK ideas with conventional practice were identified as pathways for improvement.

Also, to provide for sustainability of mentorship process in participating schools, MASTER staff recruited, in consultation with school leaders, the most capable and committed teachers in each school to serve as new mentors for residents. The selected mentors engaged in learning research-tested mentoring techniques over the school year, in preparation for being matched with residents in their disciplines and initiating 5 days of co-planning and other activities with their assigned residents in summer.

Induction Year Coaching NYSCI and Hunter College staff codesigned and implemented an induction year coaching position to provide induction year support for resident teachers in the program. There were monthly check-ins with the coaches to learn about resident teacher progress and challenges and to provide professional support for coaches. Induction coaches also regularly checked-in with host school administrators to understand how the administration viewed the inductees progress and if any problems were evident in their practice from the school leadership perspective. In this model, the coach observed the resident teacher twice a semester, which is the Hunter requirement. Optional support hours were then budgeted (a total of about one extra observation per resident teacher) for each semester.

The coach used extra hours to provide support for resident teachers who were struggling or who needed additional help in particular areas of their practice. This support was particularly beneficial for those who were struggling in the induction year. To ensure the assistance to all the resident teachers was target, and the plan for year 2 was to be clearer about expectations for and purposes of the extra coaching support. NYSCI also continued to provide support for changes in curricula, pedagogy, assessment, and integration of workshop and clinical experiences for STEM teacher candidates and provided voluntary Defense of Learning "office hours" designed and run by the NYSCI science coaches to help residents improve teacher inquiry. But residents did not widely use extra support in practice. Office hours proved less effective than support that was already integrated into monthly seminars to support DOL teacher inquiry work and discussions in the coaches' monthly meeting about residents' DOL support. Being sensitive to time management issues around the variety of support elements, helping residents recognize their needs, and discriminating about what kind of support they need, helped them to adequately clarify and more optimally use support resources.

12.5.3 Assistance with Practice-Focused Coursework

The goal of this project component was to collaborate with Hunter College in providing coherence between the reality of clinical practice and course content in order to inject PCK into graduate coursework. Close collaboration between Hunter faculty and NYSCI curriculum design resources helped identify and bridge potential gaps among fieldwork, instruction, and the clinical experience. NYSCI provided expertise in developing syllabi and assisting with instruction, providing observation and feedback on teaching, helping design an assessment rubric geared toward assessing science inquiry instruction, and facilitating integration of field experience, instructional methods, and teaching for residents.

Methods Course In the first summer of MASTER, the NYSCI instructor taught the methods course, which allowed it to be easily coordinated with the NYSCI field experience. The methods course includes analysis of student thinking/learning interactions from NYSCI fieldwork, team collaboration, and the use of learning progressions in development and revisions of units of instruction, readings about misconceptions and student learning of disciplinary practices, and working with NGSS.

Bio PCK NYSCI collaborated with Hunter to redesign the disciplinary course in biology to squarely focus it on the pedagogical content knowledge central to the teaching of biology. A bench scientist and science educator collaboratively designed the course. The NYSCI instructor worked closely with the Hunter Bio PCK instructor to develop the rubric and assess and norm final projects, particularly during the first quarter of the semester as well as providing video recording of instruction for review. The Bio PCK course incorporated field experience into weekly coursework as well as into final course project. It included collecting evidence of student thinking on a topic and assessment of that thinking in accordance to learning progression and class discussion about that evidence. The course included recognizing misconceptions, developing learning progressions, developing probes to surface student misconceptions, and analyzing student thinking on specific topics using the learning progressions they developed. Residents were required to write a final paper defining and articulating high-level understandings of a concept and refining of a learning progression, analysis of student thinking they gathered using the refined learning progression, and the implications for instruction. Developing learning progressions proved the most challenging aspect of the course development for the following reasons:

- Some weekly topics lacked adequate misconceptions literature.
- Students in the course did not have consistent enough clinical time in to collect evidence of student thinking to validate learning progressions.
- Weekly format for units of instruction did not allow enough flexibility to help residents see how concepts connect in biology.
- Learning curve for residents in gathering evidence and devising appropriate probes of student knowledge was too steep and required strong facilitation and support.

PCK in Chemistry Content in the PCK chemistry course during the fall semester focused on content discussions, assessment of residents' conceptual understanding, and what understanding of key concepts is appropriate for a high school student. Data on student thinking was collected, analyzed, and discussed semiweekly with respect to particular core concepts in chemistry (e.g., particulate nature of matter) and the concepts that hindered development or contributed to misconceptions. Probes were developed for student thinking on these key concepts and misconceptions readings in chemistry were assigned. Final projects included explication of a correct understanding of a key concept, analysis of student thinking throughout the semester, and implications for instruction. The NYSCI science instructor provided minimal support and video recording of instruction for review along with assistance in developing the rubric and assessments of final projects.

The spring semester chemistry course used a more interactive/seminar style approach to core concepts in chemistry as a model for how to think about other areas of science. For instance, a concept map of how topics are connected to major concepts in physical science was used along with misconceptions readings about major chemistry topics in physical science. Instructors collected data on student thinking about chemistry ideas (e.g., molecular bonding). There were then opportunities for resident to develop and enact instructional activities around these core chemistry ideas. The interdisciplinary nature of the class placed emphasis on the chemistry domain, which the instructor is most familiar with. The NYSCI science instructor provided development of these resources and others for Physical Science instruction based on the development of the spring chemistry course.

Practicum The practicum was intended to address classroom activity that features higher-order thinking in science and relevance of instruction to Common Core State Standards. The practicum included a lesson planning assignment (lesson plan and reflection on enactment), a differentiated classroom assignment, a formative assessment project, and an instruction on classroom management and philosophy. The NYSCI instructor facilitated collaboration between the field supervisor and practicum instructor to address resident needs and to align supports to helping residents meet the Hunter observation rubric expectations. Part of this requirement is training on and use of the School of Education Video Analysis of Teaching (VAT) system, which allows online recording and editing of teaching practice for reflection and analysis. Residents were required to submit a video recording of a full class period for the field supervisor for observation purposes. The NYSCI instructor worked with New Visions and Hunter College to redesign the science residents' Defense of Learning and practicum course projects to include a focus on science literacy practices, which align with the Educators Teacher Performance Assessment (EdTPA) (SCALE 2015) requirements and the NGSS science practices (NGSS Lead States 2013, Appendix F). The NYSCI instructor also collaborated with practicum instructor in development of the syllabus and course project for the spring practicum and provide technical and course activity support for instructor and residents for the Hunter VAT system. The practicum course included class discussions about work on final projects, such as a rubric and learning progression for student development in one of the NGSS practices, development of activities to support learning of NGSS practices in a context where students are meaningfully engaging with science content, analysis of student behavior, work, and thinking with respect to that practice, and a final paper and presentation on this work that feeds into their Defense of Learning in the spring semester. It also included in-class discussions about general teaching issues that emerge in field experience (e.g., developing relationships with students, managing demands, interview tips, etc.) and occasionally in the context of teaching science (e.g., managing controversial topics and ethical issues as a science teacher). Course project involved analyzing clips of classroom interactions for student development in a science literacy skill.

12.6 Research Approach, Implementation, and Results

12.6.1 Original Research Approach and Implementation Framework

Our original research approach would draw upon the knowledge and skills of all partners and our external evaluation team to investigate mentors' and residents' levels of efficacy and effectiveness before, during, and after their professional development experiences. Using Hunter and NYSCI's video expertise and New Visions' assessment expertise, we planned to videotape mentors and residents and use protocols set forth in the MET study (Kane et al. 2012) to document both initial practice and shifts in practice—and the impact of those shifts on student learning. One of the four observational tools in the MET study (the Danielson Framework for Teaching) is currently in use in the city to evaluate teacher practice, giving MASTER the opportunity to compare data on project participants to that of other New York City science and mathematics teachers. We also assess mentors' and residents' PCK in mathematics (Ball et al. 2008b), and, using the tools Ball et al. describe as a model, we planned to develop a set of tools for science teachers, drawing on current research on science teachers' PCK perceptions (Pell and Jarvis 2003; Barnett and Hodson 2001b) in relation to their practices and the performance of their students on standardized science tests. Importantly we intended to use teachers' own reflections on their practice, their perceptions of science or mathematics as a discipline, and their beliefs about what counts as effective science and mathematics teaching, to further inform the mentor professional development curricula and refine existing coaching models.

The questions posed as part of our internal research and development, and the specific tools and data sources to be employed to explore each one, included:

- What is the baseline content knowledge and PCK in mathematics and science of mentors and residents (existing PCK assessments in mathematics and science assessment developed as part of the project and appropriate measures of content knowledge; praxis scores, other program admission criteria, including undergraduate content degrees, GPAs of residents)?
- To what extent do residents' and mentors' sense of efficacy and effectiveness improve as a result of their professional development experiences (pre-/post-data from survey tools developed by and modeled after Ball et al.; process feedback from external evaluation)?
- What observable changes take place in preservice, early career, and experienced teachers' classrooms (periodic data from New Visions suite of assessment tools, especially lesson design, and Danielson rubrics gauging changes in levels of students' engagement and exploration of their own and classmates' thinking on conceptual questions; video analysis)?
- To what extent are changes in teacher outcomes and student achievement outcomes mediated by demographic factors such as grade and years of teaching experience, by motivational or affective characteristics such as teaching efficacy and science and mathematics attitudes, and by characteristics of implementation such as the quality of student inquiry (Danielson data, internal survey data, mentors' and residents' formative assessment data, external analysis of student achievement data)?
- Are there differential program benefits for subgroups of students and for those who have had more and less exposure to participating teachers (Danielson data, internal survey data, mentors' and residents' formative assessment data, external analysis of student achievement data)?

Our initial hypotheses were that the new model would result in a stronger corps of teachers and improved student achievement and interest in mathematics and science. The research design was based on a theory of change (Connell and Kubisch 1998) and logic model that would have been refined in year 1 and revisited throughout the project, in collaboration with the external evaluators, who have previously explored, with the project partners, the logic of the program components and their hypothesized relationships (Dede and Rockman 2007; McLaughlin and Jordan 1999). This periodic review of the logic model was intended to not only help the external evaluation team gauge progress toward goals and outcomes but also identify successful activities and "critical components" of the project that could be generalized as others explore and implement the model.

12.6.2 Implementation

In actual practice the research framework was significantly modified due to a number of factors, in particular, highly variable n in each cohort, which confounded the reliable use of quantitative measures. Emphasis was therefore placed on qualitative measures, which were used to draw conclusions and iteratively modify the program. They included a variety of formative and summative assessment methodologies. The research instruments included observation of instruction, Defense of Learning measures, open-ended surveys, and research reviews with residents, mentors, and coaches.

Project partners engaged in individual and collaborative research efforts, guided by what teachers with effective PCK skills do. New Visions conducted a study to explore the data captured by its suite of resident assessment tools, identifying elements in each tool's rubric that aligned with the learning objectives, and looked specifically at indicators of PCK-related practices captured in the NYSCI coach's use of the Danielson ratings. This review revealed some redundancy in the objectives and resulted in revisions of the objectives with the NYSCI coach. Then, New Visions created a crosswalk with the Danielson domains and conducted a series of reviews to analyze mentors' ratings of residents to explore how, and how frequently they refer to PCK and whether more frequent, explicit references to PCK were linked to residents' performance (NGSS Lead States 2013). NYSCI and New Visions also used the Defense of Learning as a measure of the efficacy of coaching strategies that impact science resident development of instructional decision-making based on student thinking and addressing misconceptions or disciplinary skill gaps and their ability to use formative assessment to continually evaluate and monitor student understanding of content, concepts, and practices. NYSCI developed a survey to measure residents' views on the nature of science/math, the nature of teaching and learning in science/math, and their ideas and views on disciplinary modeling. This survey was developed from a collection of previously validated surveys on disciplinary modeling in the nature of science and math. While the survey could not be fully implemented due to Institutional Review Board limitations, it indicated the value and need for a structured protocol for attitudes and prior knowledge about the structure of science ideas and processes that are not elicited by other means. NYSCI also developed an approach to study teacher noticing of student learning as a way that practice can be improved using a variety of inputs including: resident video recordings to reflect on their practice for the Hunter VAT requirement and edTPA portfolio, residency coaching work, and Defense of Learning data. Ultimately, this work focused on how experience in an informal institution (NYSCI) influenced the resident's disciplinary teaching practice development.

In terms of coursework, maintaining the quality of instruction for PCK over the course of the program was a significant challenge. For some disciplines, such as biology, the same instructor taught the course for multiple years, but for others, instructors changed annually. Many courses were instructed by adjuncts, who were not required to closely collaborate with the project personnel or attend the same kind of professional development as full-time tenured faculty. While NYSCI and New Visions residency staff collaborated with Hunter faculty to design new and innovative courses whenever possible, how the courses were conducted was ultimately tied to specific teacher educators, and there was no consistency from year to year.

12.6.3   Effect of Mentorship on Practice

Mentors regularly reported positive feedback on PCK professional development activities and displayed more ownership over the conversation as a result of the program. The following are some quotes from mentors about the professional development and the effect of mentoring on their practice in response to the prompt "I used to... but now I..." :

- *"I used to give [students] the right answer, but now I let them struggle for a while, reflect on the problem, because it helps the students make sense of it and focus more on the process."*
- *"I am hyper-aware of student responses in whole class discussions and small groups because of watching how [the resident] responds to student misconceptions."*
- *"I used to not look at student work in depth. but now I look at it systematically because I have seen that it is useful for informing teaching practice."*
- *"I used to summarize learning for the students, but now I require students to articulate big ideas or big questions for themselves because it allows me to assess their progress."*
- *"I used to move through a lesson without assessing students' understandings in a clear way until close to the ending of the lesson. But now I assess more throughout because my resident challenged me to do so on a regular basis!"*

12.6.4 Successes and Challenges to Induction Coaching

The practices of holding monthly check-ins with the coaches to learn about novice teacher progress and challenges and providing the professional support for the coaches seemed to work well in the induction coaching process. It was also evident that close communication between the induction coach and the practicum instructor about the novices' progress (and challenges) was very helpful in coordinating support for the resident teachers. An important challenge was how widely school-based support and professional culture varied by site and how this variation can hinder resident teacher growth.

12.6.5 Feedback from Residents

The intent was for the informal learning setting of NYSCI to give residents more "experiential" training in conjunction with theoretical knowledge about PCK. The intent was to provide background in PCK and inquiry-based learning so that it could inform residents' experience in the clinic and scaffold the learning of PCK in their graduate coursework. The fact that the fieldwork came before their experience in the clinic made it difficult for residents to directly apply what they learned. Residents suggested that there needed to be more support to apply the ideas in teaching practice as their competency increased. They also struggled with understanding and applying PCK theory, which seemed to demand background knowledge in teaching and learning theory that residents lacked. A similar gap was noted for project-based learning strategies, in which it was evident that novice teachers could only begin to use them once they were comfortable enough with classroom management, pacing, and routine structures and strategies to enhance them with project-based or inquiry work. It wasn't that residents did not see the value in less structured or more abstract activities, nor is it a matter of whether certain training or coursework was helpful to develop and deepen PCK; it had more to do with having the mental models and knowledge structures in which to fit this very different knowledge (Sloan et al. 2016).

Fieldwork at NYSCI was subsequently revised to train new teachers to use certain professional norms (e.g., evidence-based reflection on, and analysis of teaching, design-based and disciplinary modeling instructional work, developing learning objectives, and introducing them to innovative instructional approaches to help them set the path to creative and thoughtful pedagogy in the clinical experience). This was complemented by coordination with Hunter College Methods I instructors. Observation and feedback from participants indicated improvement in identifying and characterizing good learning objectives and how to connect them to standards and that there was a deeper understanding of how to define, circumscribe, and apply modeling and design in classroom practice, something that was indicated as problematic in the previous year's field experience.

In the final year of summer field experience NYSCI took a slightly different approach based on the need to focus more on applying knowledge and understanding integration of PCK into instruction. The first phase was "teachers as students," the second was "teachers as teachers," and the third was "teachers as curriculum designers." Having residents see their development activities through these different lenses affected how residents thought about their instructional activities. By the end of the workshops, there was a noticeable change in their perception of the teaching profession and how much they have to learn to improve their teaching practices. Residents learned about science content, lesson design, and enactment and began to develop their capacities as reflective practitioners through a review of their own video recordings.

12.7 Conclusions and Recommendations

As described at the outset of this chapter, pedagogical content knowledge is a particularly complex construct demanding a significant theoretical foundation to be well understood. Integration of PCK into preservice teacher development and, further, into clinical practice is, thus, demanding both theoretically and in developing, executing, and assessing learning. The MASTER program was structured to intentionally, directly, and meaningfully integrate PCK into a preservice learning program for graduate high school science and math education practice. MASTER provided a useful test bed for bringing together the very different lenses on learning provided by high-needs public schools, a public graduate school of education, and a cultural institution. As residency programs expand and become more commonplace, we expect there will be increased interest in using PCK as a framing for learning. Recognizing the complexity of the clinical learning environment, the demanding role of practitioners in facilitating learning in more sophisticated ways, and the value of collaboration with informal STEM institutions in bringing formative, inquiry-based teaching has the potential for a renaissance in teaching and learning.

Results of this program as gauged through a variety of formative and summative assessment mechanisms indicate that the practices developed through MASTER have the potential to be replicable, extensible, useful, and transformative. It indicates the power of residency models for changing the way educators are prepared for practice. We have assembled a series of recommendations that follows to help understand how many of the programs challenges could be addressed.

12.7.1 Induction Coaching Practices

We recommend that induction coaches use the following practices:

- Hold monthly check-ins with the coaches to learn about novice teacher progress and challenges.

- Provide the professional support for the coaches seem to work well in the induction coaching process.
- Foster close communication between induction coaches and practicum instructors to coordinate support for the resident teachers.

12.7.2 Effective Support Mechanisms for Residents

To provide residents background in PCK and inquiry-based learning from an informal science perspective to their experience in the clinic and scaffold the learning of PCK in their graduate coursework, it is critical to attend to the following recommendations:

- Avoid offering the fieldwork before their experience in the clinic.
- Provide more support to apply the ideas in teaching practice.
- Offer them background knowledge in teaching and learning theory to ensure they are understanding and applying PCK theory.
- Offer them classroom management, pacing, and routine structures and strategies to enhance their comfort with the implementation of project-based or inquiry work.
- Be sensitive to time management issues around the variety of support elements, help residents recognize their needs, and be discriminating about what kind of support they need.
- Train them to use evidence-based reflection for analysis of teaching, design-based and disciplinary modeling of instructional work, and developing learning objectives.
- Focus the field experience at the informal science institution on applying knowledge and understanding integration of PCK into the development of instructional activities through three different lenses: "teachers as students," "teachers as teachers," and "teachers as curriculum designers."

12.7.3 PCK Research Reviews

To identify elements in each tool's rubric that aligned with the list of learning objectives and look specifically at indicators of PCK-related practices captured in coach's use of the Danielson ratings, it is important to engage in the following activities:

- Review learning objectives periodically.
- Create a crosswalk with the Danielson domains, and analyze mentors' ratings of residents to explore how, and how frequently they refer to PCK and whether more frequent, explicit references to PCK are linked to residents' performance.

- Employ the Defense of Learning as a measure of the efficacy of coaching strategies that impact science resident development and their ability to use formative assessment.
- Use data from video recordings, residency coaching work, and Defense of Learning to determine how experience in an informal institution (in this case, NYSCI) can influence resident's disciplinary teaching practice development.

12.7.4 Quality and Consistency of Coursework

To maintain the quality of instruction for PCK and consistency from year to year, the following activities must be taken into account:

- Recruit full-time tenured faculty who are accountable to professional development needs of the program.
- Work with the same course instructor whenever possible.
- Require close collaboration among project personnel.
- Ensure that all instructors attend the same kind of professional development.

Unique to the use of PCK in science teaching is the conclusion that, while all the science disciplines overlap one another, they each constitute a different lens on nature and require substantial content knowledge for competent teaching practice. While residents in MASTER were selected from undergraduate programs in the sciences, PCK puts particular demands on both breadth and depth of knowledge and scaffolds that knowledge with pedagogical strategies and tactics that challenge even the most content-knowledgeable candidates. But the complexity and demanding nature of PCK in practice also requires strong collaboration among experts, instructors, curriculum specialists, and intensive mentoring and coaching, as well as ambitious and thoughtful practitioners. The success of PCK integration depends heavily on the capacity for these entities to collaborate well, receive adequate support and training, and tightly integrate PCK practices into the complex and dynamic clinical environment. Over the course of the MASTER project, these capacities for PCK development in residents were improved incrementally, but discoveries made along the way indicate there is much more to be done.

References

Anwari, I., Yamada, S., Unno, M., Saito, T., Suwarma, I., Mutakinati, L. A., & Kumano, Y. (2015). Implementation of authentic learning and assessment through STEM education approach to improve students' metacognitive skills. *K-12 STEM Education, 1*(3). Bangkok: Institute for the Promotion of Teaching Science and Technology, 123.

Autor, D., Levy, F., & Murnane, R. (2003). The skill content of recent technological change: An empirical exploration. *Quarterly Journal of Economics, 118*(4). Cambridge, MA: MIT Press, 1279.

Ball, D., & Cohen, D. (1999). Developing practice, developing practitioners: Toward a practice-based theory of professional education. In L. Darling-Hammond & G. Sykes (Eds.), *Teaching as the learning profession: Handbook of policy and practice* (p. 3). San Francisco: Jossey-Bass.

Ball, D., Thames, M., & Phelps, G. (2008a). Content knowledge for teaching: What makes it special? *Journal of Teacher Education, 59*(5). Thousand Oaks: Sage Publications, 389.

Ball, D., Thames, M., & Phelps, G. (2008b). Content knowledge for teaching: What makes it special? *Journal of Teacher Education, 59*(5), 389–407.

Barnett, J., & Hodson, D. (2001a). Pedagogical context knowledge: Toward a fuller understanding of what good science teachers know. *Science Education, 85* . New York: Wiley, 426. https://doi.org/10.1002/sce.1017.

Barnett, J., & Hodson, D. (2001b). Pedagogical content knowledge: Toward a fuller understanding of what good science teachers know. *Science Education, 85*, 426–453.

Berry, B., Montgomery, D., Rachel, C., Hernandez, M., Wurtzel, J., & Snyder, J. (2009). *Creating and sustaining urban teacher residencies: A new way to recruit, prepare, and retain effective teachers in high-needs districts. Report*. Washington, DC: Aspen Institute.

Black, P., & Wiliam, D. (2009). Developing the theory of formative assessment. *Educational Assessment, Evaluation, and Accountability, 21*(1). Dordrecht: Springer, 5.

Black, P., Harrison, C., Lee, C., Marshall, B., & William, D. (2004). *Assessment for learning-putting it into practice*. Maidenhead: Open University Press.

Bransford, J., Brown, A., & Cocking, R. (1999). *How people learn*. Washington, DC: National Academy Press.

Brown, J., Collins, A., & Duguid, P. (1989). Situated cognition and the culture of learning. *Educational Researcher, 18*(1). Thousand Oaks, CA: Sage Publications, 32.

Chin, C. (2004). Museum experience – A resource for science teacher education. *International Journal of Science and Mathematics Education, 2*. Dordrecht: Springer, 63.

Chubin, D., May, G., & Babco, E. (2005). Diversifying the engineering workforce. *Journal of Engineering Education, 94*, 73.

Connell, J., & Kubisch, A. (1998). Applying a theory of change approach to the evaluation of comprehensive community initiatives: Progress, prospects, and problems. In K. Fullbright Anderson, A. Kubisch, & J. Connell (Eds.), *New approaches to evaluating community initiatives* (Vol. 2, pp. 15–44). Washington, DC: The Aspen Institute.

Costa, A., & Garmston, R. (1994). *Cognitive coaching: A foundation for renaissance schools*. Norwood, MA: Christopher-Cordon Publishers, Inc.

Cross, J. (2007). *Informal learning: Rediscovering the natural pathways that inspire innovation and performance*. San Francisco: Pfeiffer.

Dana, N., & Yendol-Hoppey, D. (2009). *The reflective educator's guide to classroom research* (2nd ed.). Thousand Oaks: Corwin Press.

Danielson, C. (2013). *The framework for teaching evaluation instrument*. Princeton, NJ: The Danielson Group.

Dede, C., & Rockman, S. (2007). Lessons learned from studying how innovations can achieve scale. *Threshold Magazine, Spring 2007*, 4–10.

Dewey, J. (1902). *The school as a social centre. The middle works of John Dewey* (Vol. 2, pp. 80–96).

Dikilitaş, K. (2016). *Innovative professional development methods and strategies for STEM education* (pp. 1–311). Hershey, PA: IGI Global. https://doi.org/10.4018/978-1-4666-9471-2.

Doppelt, Y., & Schunn, C. D. (2008). Identifying students' perceptions of the important classroom features affecting learning aspects of a design-based learning environment. *Learning Environment Research, 11*(3 . Dordrecht: Springer, 195). https://doi.org/10.1007/s10984-008-9047-2.

van Driel, J., & Berry, A. (2012). Teacher professional development focusing on pedagogical content knowledge. *Educational Researcher, 41*(1). Thousand Oaks, CA: Sage Publications, 26.

Edelson, D. (2001). Learning-for-use: A framework for the design of technology-supported inquiry activities. *Journal for Research in Science Teaching, 38* . New York: Wiley & Sons, 355. https://doi.org/10.1002/1098-2736(200103)38:3<355::AID-TEA1010>3.0.CO;2-M.

Evans, K., Hodkinson, P., Rainbird, H., & Unwin, L. (2006). Improving workplace learning. In R. Millar, J. Leach, J. Osborne, & M. Ratcliffe (Eds.), *Improving subject teaching: Lessons from research in science education*. London: Routledge.

Fortus, D., Dershimer, R., Krajcik, J., Marx, R., & Mamlok-Naaman, R. (2004). Design-based science and student learning. *Journal of Research in Science Teaching, 41* . New York: Wiley & Sons, 1081. https://doi.org/10.1002/tea.20040.

Gardner, H. (1991). *The unschooled mind*. New York: Basic Books.

Geier, R., Blumenfeld, P. C., Marx, R. W., Krajcik, J. S., Fishman, B., Soloway, E., & Clay-Chambers, J. (2008). Standardized test outcomes for students engaged in inquiry-based science curricula in the context of urban reform. *Journal for Research in Science Teaching, 45* . New York: Wiley & Sons, 922. https://doi.org/10.1002/tea.20248.

Gerber, B., Cavallo, A., & Marek, E. (2001). Relationships among informal learning environments, teaching procedures and scientific reasoning ability. *International Journal of Science Education, 23*(5). London: Taylor and Francis.

Gess-Newsome, J., & Lederman, N. (2001). *Examining pedagogical content knowledge: The construct and its implications for science education*. Dordrecht: Springer Science & Business Media.

Graham, C. (2005). Blended learning systems: Definition, current trends, and future directions. In C. Bonk & C. Graham (Eds.), *Handbook of blended learning: Global perspectives, local designs*. San Francisco, CA: Pfeiffer Publishing.

Griffin, J. (1994). Learning to learn in informal science settings. *Research in Science Education, 24* . Dordrecht: Springer, 121. https://doi.org/10.1007/BF02356336.

Grossman, P., Schoenfeld, A., & Lee, C. (2005). Teaching subject matter. In L. Darling-Hammond & J. Bransford (Eds.), *Preparing teachers for a changing world: What teachers should learn and be able to do* (p. 201). San Francisco: Jossey Bass.

Hashweh, M. (2005). Teacher pedagogical constructions: A reconfiguration of pedagogical content knowledge. *Teachers and Teaching, 11*(3). Taylor and Francis.

Honey, M., & Kanter, D. (2013). *Design, make, play: Growing the next generation of STEM innovators*. New York: Routledge.

Jung, M., & Tonso, K. (2006). Elementary pre-service teachers learning to teach science in science museums and nature centers: A novel program's impact on science knowledge, science pedagogy, and confidence teaching. *Journal of Elementary Science Education, 18* . Dordrecht: Springer, 15. https://doi.org/10.1007/BF03170651.

Kane, T., McCaffrey, D, Miller, T, and Staiger, D. (2012). Have we identified effective teachers? Validating measures of effective teaching using random assignment http://k12education.gatesfoundation.org/download/?Num=2676&filename=MET_Validating_Using_Random_Assignment_Research_Paper.pdf. Accessed 8 June 18.

Kirshner, D., & Whitson, J. (Eds.). (1997). *Situated cognition: Social, semiotic, and psychological perspectives*. Mahwah, NJ: Lawrence Erlbaum Associates, Inc.

Klein, E., Taylor, M., Onore, C., Strom, K., & Abrams, L. (2013). Finding a third space in teacher education: Creating an urban teacher residency. *Teaching Education, 24*(1). London: Taylor and Francis.

Kolodner, J., Crismond, D., Gray, J., Holbrook, J., & Puntambekar, S. (1998). Learning by Design from theory to practice. In *Proceedings of the International Conference of the Learning Sciences (ICLS 98)* (pp. 16–22). Charlottesville, VA: AACE.

Leach, J., Scott, P., Ametller, J., Hind, A., & Lewis, J. (2006). Implementing and evaluating teaching interventions: Towards research evidence-based practice? In R. Millar, J. Leach, J. Osborne, & M. Ratcliffe (Eds.), *Improving subject teaching: Lessons from research in science education*. London: Routledge.

Lederman, N., Abd-El-Khalick, F., Bell, R., & Schwartz, R. (2002). Views of nature of science questionnaire: Toward valid and meaningful assessment of learners' conceptions of nature of science. *Journal of Research in Science Teaching, 39* . New York: Wiley, 497. https://doi.org/10.1002/tea.10034.

Lee, O. (2004). Teacher change in beliefs and practices in science and literacy instruction. *Journal of Research in Science Teaching, 41*(1). New York: Wiley, 93.

Loucks-Horsley, S., Stiles, K., Mundry, S., & Hewson, P. (2009). *Designing professional development for teachers of science and mathematics*. Thousand Oaks: Corwin Press, Sage.

Loughran, J., Mulhall, P., & Berry, A. (2004). In search of pedagogical content knowledge in science: Developing ways of articulating and documenting professional practice. *Journal for Research in Science Teaching, 41* . New York: Wiley, 370. https://doi.org/10.1002/tea.20007.

McLaughlin, J., & Jordan, G. (1999). Logic models: A tool for telling your program's performance story. *Evaluation and Program Planning, 22*(1), 65–72.

Mehalik, M., Doppelt, Y., & Schuun, C. (2008). Middle-school science through design-based learning versus scripted inquiry: Better overall science concept learning and equity gap reduction. *Journal of Engineering Education, 97* . New York: Wiley & Sons, 71. https://doi.org/10.1002/j.2168-9830.2008.tb00955.x.

Millar, R., & Hames, V. (2006). Using designed teaching materials to stimulate changes in practice. In R. Millar, J. Leach, J. Osborne, & M. Ratcliffe (Eds.), *Improving subject teaching: Lessons from research in science education*. London: Routledge.

Moje, E., Collazo, T., Carrillo, R., & Marx, R. (2001). "Maestro, what is 'quality'?": Language, literacy, and discourse in project-based science. *Journal of Research in Science Teaching, 38*(4). New York: Wiley, 469.

National Center for Teacher Residencies. (2016). A case study of the MASTER teacher residency: Building novice STEM teacher capacity through innovative pedagogies and coursework. In *The case study project* (Vol. 8). Chicago, IL: National Center for Teacher Residencies.

National Governors Association Center for Best Practices. (2010). *Common core state standards*. Washington, DC: National Governors Association Center for Best Practices, Council of Chief State School Officers.

NGSS Lead States. (2013). *Next generation science standards: For states, by states*. Washington, DC: The National Academies Press.

Pacific Policy Research Center. (2010). *21st century skills for students and teachers*. Honolulu: Kamehameha Schools, Research & Evaluation Division.

Park, S., & Oliver, S. (2008a). Revisiting the conceptualisation of pedagogical content knowledge (PCK): PCK as a conceptual tool to understand teachers as professionals. *Research in Science Education, 38*(3). Dordrecht: Springer, 261.

Park, S., & Oliver, J. (2008b). *Revisiting the conceptualisation of pedagogical content knowledge (PCK): PCK as a conceptual tool to understand teachers as professionals, Research in Science Education* (Vol. 38, p. 261). Dordrecht: Springer. https://doi.org/10.1007/s11165-007-9049-6.

Pell, A., & Jarvis, T. (2003). Developing attitude to science education scales for use with primary teachers. *International Journal of Science Education, 25*(10), 1273–1295.

Pella, S. (2015) Pedagogical reasoning and action: Affordances of practice-based teacher professional development.

Remillard, J., & Geist, P. (2002). Supporting teachers' professional learning by navigating openings in the curriculum. *Journal of Mathematics Teacher Education, 5*(1). Dordrecht: Springer, 7.

Rinke, C., Mawhinney, L., & Park, G. (2014). The apprenticeship of observation in career contexts: A typology for the role of modeling in teachers' career paths. *Teachers and Teaching, 20*(1). London: Taylor and Francis.

Saxman, L., Gupta, P., & Steinberg, R. (2010). CLUSTER: University-science center partnership for science teacher preparation. *The New Educator, 6*. London: Taylor and Francis, 280.

SCALE. (2015). *Review of research on teacher education: edTPA task dimensions and rubric constructs*. Stanford, CA: Stanford Center for Assessment, Learning and Equity.

Seymour Papert, S., & Idit Harel, I. (1991). *Constructionism*. New York: Ablex Publishing Corporation.

Shepard, L. (1989). Why we need better assessments. *Educational Leadership, 46*(7). Washington, DC: Association for Supervision and Curriculum Development, 4.

Shulman, L. (1986). Those who understand: Knowledge growth in teaching. *Educational Researcher, 15*(2). Thousand Oaks: Sage Publications, 4.

Sloan, K., Allen, A., Bass, K., Blazevski, B., & Li, J. (2016). The MASTER model: Linking theory, preparation, practice, and performance. In *Year 3 evaluation report for the mathematics and science teacher education residency program*. San Francisco, CA: Rockman, et al..

Solomon, J. (2009). The Boston teacher residency: District-based teacher education. *Journal of Teacher Education, 60*(5). Thousand Oaks, CA: Sage Publications, 478.

Sondergeld, T., Peters-Burton, E., & Johnson, C. (2016). Integrating the three dimensions of next generation science standards: Issues and solutions for authentic assessment of student learning. *Journal of School Science & Mathematics, 116* . New York: Wiley, 67. https://doi.org/10.1111/ssm.12160.

Starkey, L. (2010). Teachers' pedagogical reasoning and action in the digital age. *Teachers and Teaching, 16*(2), 233.

Thirunarayanan, M. (1997). Promoting pre-service science teachers' awareness of community-based science education resource centers. *Journal of Science Teacher Education, 8* . Dordrecht: Springer, 69. https://doi.org/10.1023/A:1009405619969.

Tuttle, N., Kaderavek, J., Molitor, S., Czerniak, C., Johnson-Whitt, E., Bloomquist, D., Namatovu, W., & Wilson, G. (2016). Investigating the impact of NGSS-aligned professional development on PreK-3 teachers' science content knowledge and pedagogy. *Journal of Science Teacher Education, 27*(7). Dordrecht: Springer.

Wilkes, R. (1994). *Using Shulman's model of pedagogical reasoning and action in a pre-service program*. Paper presented at the Annual Meeting of the Australian Teacher Education Association. 24th, Brisbane, Quensland, Australia, July 3–6,1994.

Chapter 13
Developing Educative Materials to Support Middle-School Science Teachers' PCK for Argumentation: Comparing Multimedia to Text-Based Supports

Suzanna Loper, Katherine L. McNeill, and Megan Goss

Abstract Science teachers need support in developing pedagogical content knowledge (PCK) around science practices such as argumentation. Educative curriculum materials (ECMs) have great potential as a tool to support teachers' PCK development, particularly for in-service teachers. In this chapter, we describe an intervention in which we developed and researched web-based educative curriculum materials focused on supporting teachers' PCK of argumentation. Two different versions of the materials—one with text-based educative supports and one with text-based supports plus multimedia elements—were tested with a total of 90 middle-school science teachers. Participants completed pre- and post-surveys designed to measure their PCK of argumentation. In addition, back-end data was collected on teachers' use of the web-based materials. Findings indicated that teachers' PCK did increase from pre to post, although there was no significant difference between the two versions of the curriculum. Teachers' use of the materials varied widely. Our findings suggest that educative curriculum materials are a promising avenue for supporting teacher PCK. Future efforts should take into consideration teachers' use patterns in order to increase the likelihood that teachers will use and benefit from educative elements.

Keywords PCK · Scientific argumentation · Teacher learning · Middle-school teaching · Educative curriculum · Multimedia

S. Loper (✉) · M. Goss
Lawrence Hall of Science, Berkeley, CA, USA

K. L. McNeill
Boston College, Boston, MA, USA

© Springer International Publishing AG, part of Springer Nature 2018   241
S. M. Uzzo et al. (eds.), *Pedagogical Content Knowledge in STEM*,
Advances in STEM Education, https://doi.org/10.1007/978-3-319-97475-0_13

13.1   Introduction

Current science education reform efforts in the United States are creating a greater
need to support science teacher learning. The Next Generation Science Standards,
adopted in 18 states and influencing new standards in many other states, include
a strong focus on science practices and require that teachers engage students in
three-dimensional learning, which integrates the "three dimensions" of (1) science and engineering practices, (2) disciplinary core ideas, and (3) crosscutting
concepts (NGSS Lead States 2013; Pruitt 2014). This vision of science education
goes far beyond previous standards that separated science content from "inquiry
skills" and thus presents new challenges and learning demands for teachers. The
research and development described in this chapter focuses on supporting teachers in teaching the science practice of *arguing from evidence*. Specifically, we
focus on teachers' pedagogical content knowledge (PCK) for scientific argumentation. Research on supporting PCK for scientific argumentation is still limited
(McNeill et al. 2016a; McNeill and Knight 2013; Zohar 2008), and the work that
has been done has been primarily in the context of pre-service educators (e.g.,
Avraamidou and Zembal-Saul 2005; Hume and Berry 2011; Osana and Seymour
2004; Zembal-Saul et al. 2002) or in professional development with in-service
teachers (e.g., McNeill and Knight 2013; Osborne et al. 2004; Simon et al. 2006).
The project described in this chapter contributes to the field by investigating
whether in-service science educators' PCK can be supported through educative
curriculum materials, in particular multimedia educative curriculum materials
focused on PCK of scientific argumentation. We investigate these questions:

1. How do teachers use educative materials designed to support PCK of
   argumentation?
2. Do educative curriculum materials impact teachers' PCK of scientific
   argumentation?
3. Does the addition of multimedia elements to educative curriculum materials with
   text-based supports result in a different impact on teachers' PCK of scientific
   argumentation?

In this chapter, we describe the characteristics of the educative curriculum
materials and report results from an investigation of the impact of two different versions of the materials, with and without multimedia elements, on teachers' PCK.

13.2   Theoretical Framework

In this section, we describe theoretical perspectives on scientific argumentation,
PCK for scientific argumentation, and educative curriculum materials that form the
foundation of our research and development.

13.2.1   Scientific Argumentation

Engaging in argument from evidence is one of the eight science and engineering practices identified in the Framework for K-12 Science Education (National Research Council 2012). This practice is focused on the construction, critique, and revision of knowledge claims using evidence (Osborne 2010). In considering how scientific argumentation can be supported in the classroom, it is important to consider two aspects of argumentation: that scientific arguments have a particular structure and that scientific argumentation is a social or dialogic process, in which the scientific community interacts to support, challenge, and revise claims (Jiménez-Aleixandre and Erduran 2008). In terms of the structural aspects of a scientific argument, we use a model similar to other researchers in which an argument has a claim that answers a question about the natural world, evidence that supports that claim, and reasoning that shows how the evidence is connected to the claim (McNeill et al. 2006; Sampson and Clark 2008; Toulmin 1958). With respect to the dialogic aspects of argumentation, we focus on two components: (1) the idea that competing claims must be involved and (2) the idea that argumentation should be an interactive process. Scientific argumentation is a departure from typical science classroom instruction (Pruitt 2014), and a number of studies have documented difficulties that teachers and students have with both the structural and dialogic aspects of argumentation (Berland and Reiser 2011; McNeill and Knight 2013; Sampson and Blanchard 2012). These are described further in the next section.

13.2.2   Pedagogical Content Knowledge for Scientific Argumentation

Pedagogical content knowledge has been conceptualized as professional knowledge specific to teaching and learning about a topic (Abell 2008; Kind 2009; Shulman 1986). We have focused on two elements of PCK that are in Shulman's original conception (1986) and have also been used by numerous science education researchers (e.g., Abell 2008; Kind 2009; Magnusson et al. 1999): (1) knowledge of students' conceptions and (2) knowledge of instructional strategies.

Recent work has emphasized a dynamic view of PCK: the idea that PCK should be conceptualized as knowledge-in-use or something that a teacher *does* rather than *has* (Alonzo and Kim 2016; Cochran et al. 1993). We incorporate this perspective both in our support for PCK and our assessment of PCK. In our initial work conceptualizing PCK of scientific argumentation, one challenge that emerged was moving beyond assessing and supporting pseudoargumentation, such as surface-level features like naming argument components (e.g., claim, evidence, and reasoning), to target the quality of students' argumentation and appropriate instructional strategies (McNeill et al. 2016a). One way to address this is through considering how PCK manifests itself in action in a particular classroom context (Settlage 2013).

Table 13.1 Conceptions for PCK of argumentation (see McNeill et al. 2016a, Table 10, for more detail)

Conception #1: argumentation structure	Conception #2: argumentation as a dialogic process
Conception 1A: evidence Teachers evaluate and support students' use of high-quality evidence to justify their claims	Conception 2A: student interactions Teachers evaluate and support students in building off of and critiquing each other's ideas
Conception 1B: reasoning Teachers evaluate and support students' use of scientific ideas or principles to explain the link between their evidence and claim	Conception 2B: competing claims Teachers evaluate and support students in critiquing competing claims

Consequently, in our PCK assessment, we used specific classroom vignettes to explore teachers' PCK of argumentation. We explicitly define teachers' PCK of scientific argumentation in terms of teachers' abilities to evaluate students' engagement in argumentation and propose potential instructional strategies to address those strengths and weaknesses. This is in contrast to, for example, stating that teachers "understand evidence plays a key role in argument."

Furthermore, we apply this approach across four conceptions of argumentation that have been identified as challenging for teachers and students (McNeill et al. 2016a). We use parallel language in our four conceptions including "Teacher analyzes student discourse," "Teacher identifies any challenges," and "Teacher uses an instructional strategy." The four conceptions that we focus on (two related to argumentation structure and two related to argumentation as a dialogic process) are shown in Table 13.1.

Finally, in terms of interventions to support PCK, the bulk of work on PCK of argumentation has focused on pre-service teachers (e.g., Hume and Berry 2011; Zembal-Saul 2009) and, in some cases, work with in-service teachers in the context of professional development (e.g., McNeill and Knight 2013; Sadler 2006; Simon et al. 2006). Relatively little work has been done on the potential for curriculum materials to support PCK. Schneider and Plasman (2011) note that experience with students is an important aspect of developing PCK; because educative curriculum is used in the context of actual teaching, this may be an advantage for educative curriculum as a support for PCK.

13.2.3 Educative Curriculum Materials

Some researchers have found that PCK can be improved but argue that this requires both high-quality materials and support in the form of professional learning opportunities (Gess-Newsome et al. 2011). In this project, we were interested in the question of whether curriculum materials alone, if designed to be educative, could impact PCK. Educative curriculum materials (ECMs) are curriculum materials designed to support the learning of teachers as well as students (Ball and Cohen 1996; Davis and Krajcik 2005; Davis et al. 2014). Because teachers interact with ECMs in the context of their everyday teaching practices, they may be an efficient and scalable means to support educational reform (Davis et al. 2016). ECMs have shown some effect on teacher practice and student outcomes (Arias et al. 2016a; Cervetti et al. 2015). One important aspect of ECMs is to support the teacher as designer (Brown 2009). ECMs should support productive adaptation of curriculum materials and thus should support teachers in understanding the rationale and enacting principles of a curriculum, rather than merely following the curriculum (Davis and Krajcik 2005; McNeill et al. 2017).

As a field, we have begun to develop resources, such as curriculum materials, to provide support for teachers around argumentation (Cavagnetto 2010). However, we need more specified design principles for how to support teacher learning (Wilson 2013) as well as greater understanding of how teachers use curriculum resources to better meet their needs (Davis et al. 2016). We consider the relationship between teachers and curriculum materials to be a participatory relationship (Remillard 2005) in which characteristics of the teacher and characteristics of the curriculum materials interact to produce the enacted curriculum. Thus, consistent with other researchers (Brown 2009), we consider curriculum use to be a design activity. With this perspective, there are multiple acceptable enactments of a curriculum, rather than one correct way to teach each lesson (Remillard 2005), and successful implementation is determined by alignment with the overarching goals of the curriculum, rather than by whether teachers followed scripted procedures (McNeill et al. 2016b). Thus, one goal of educative curriculum materials is to provide rationale for instructional decisions in order to enable teachers to successfully adapt the resources to meet the needs of their particular students and context (Beyer et al. 2009).

In this project, we expanded on existing research by including a focus on multimedia educative curriculum materials, including educative video. Rich media such as video may be especially important for a practice like scientific argumentation where, because of the dialogic nature of the practice, text-based supports alone may not be sufficient (Alozie et al. 2010). Video has been used in a variety of professional learning settings, both face-to-face and online (Roth et al. 2011; van den Berg et al. 2008; Sherin 2003; Watters and Diezmann 2007). However, to date there has been limited research on video and other multimedia as part of educative curriculum materials.

13.3 Development of Intervention

Given the need for support for teacher PCK for scientific argumentation, the promise of educative curriculum materials, and the idea that multimedia materials in a web-based teacher's guide might have particular affordances for supporting

PCK for scientific argumentation, we undertook to develop and study a set of multimedia educative curriculum materials for middle-school science teachers, focused on supporting teacher learning about scientific argumentation.

13.3.1 Curriculum Context

For this study, we adapted three curriculum units that were a part of a middle-school Earth and Space Science curriculum being developed by the Lawrence Hall of Science. The materials were designed for the Next Generation Science Standards (which were in draft form at the time of the curriculum development). The curriculum had a strong focus on argumentation but also incorporated experiences around other science practices such as modeling and analyzing data. The curriculum included a total of 62 lessons across three units: *Rock Transformations (Rocks)*, *Currents and Earth's Climate (Currents)*, and *Space and Gravity (Space)*.

13.3.2 Support for PCK

The support for PCK of argumentation in the curriculum materials included two components of PCK commonly identified in the literature (Kind 2009): (1) understanding student difficulties with argumentation and (2) instructional strategies for supporting argumentation. We focused on the four conceptions described above: high-quality evidence, reasoning, student interaction, and considering competing claims.

The curriculum materials were designed to support teachers' PCK in three ways:

1. **Supports embedded in the lesson plans themselves.** The lessons, including both the student-facing materials (such as projections or student sheets) and the teacher-facing materials (such as descriptions of lesson goals and step-by-step instructions for teaching the lesson), were designed to support teachers' PCK for argumentation.
2. **Supports in text-based educative features.** Educative teacher notes, included in each lesson, provided additional support for teachers' PCK of argumentation.
3. **Supports in multimedia educative elements.** Teacher-facing short videos, slideshows, and podcasts, embedded within the lesson plans, provided additional support for teachers' PCK of argumentation.

The following are examples of how PCK of argumentation was supported in each of these aspects of the curriculum materials.

13.3.3 PCK Supports Embedded in Lesson Plans

Both the student materials and the surrounding teacher-facing content provided support for teachers' learning about argumentation, including support for understanding student difficulties with argumentation, and support for developing a repertoire of instructional strategies. Below we provide examples of how different elements of the lesson plans could support teacher PCK for argumentation.

1. **Explicit descriptions of purpose for each lesson.** One support for teacher PCK for argumentation is the explicit description of the purpose, for each lesson ("session purpose") and for each activity within a lesson ("section purpose"). In lessons focused on argumentation, these explicit descriptions of purpose help the teacher to understand the objectives related to argumentation, the challenges students might face, and the goal of the instructional activities.

   For example, in *Rocks* Lesson 2.11, the session purpose states, "Session purpose: In this final session of Investigation 2, students write a scientific argument that answers the question they have been investigating: What caused the dinosaurs to go extinct 65 million years ago? The teacher introduces the writing task, emphasizing the elements of a scientific argument, encouraging students to explain how their evidence supports their claims. Time is spent discussing language to help students do this..." In addition, a section purpose within this lesson states, "Purpose: In this section, students connect the science seminar discussion from the previous session to writing a scientific argument. They focus on how a written argument is much more than a list of evidence—it explains why the evidence is important, just as they discussed in the science seminar. This prepares students to actually write their arguments..." These explicit statements of purpose help teachers to recognize that their goal will be to help students explain why their evidence is important (one of the student challenges).

2. **Examples of instructional strategies.** Within the lesson plans, as both regular components of the lesson and optional extensions, descriptions of instructional strategies related to argumentation for the teacher are provided. Through reading about and using these strategies, the teacher could potentially build his or her repertoire of instructional strategies. For example, one instructional strategy is to provide contrasting examples to help students understand a conception such as the need to include reasoning in an argument. In *Rocks* Lesson 2.11, the lesson plan guides the teacher to project two contrasting arguments: one in which the reasoning is explained clearly, and one in which it is not. Through following lesson plans such as these, teachers could gain familiarity with instructional strategies that could be used in their teaching of argumentation. They could then use these strategies outside of the specific curriculum, thus supporting PCK of argumentation.

3. **Information about possible student thinking or student difficulties embedded within lesson steps.** In many cases, ideas related to what students might be thinking, or difficulties that students might have, are interwoven with the steps for teaching the lesson. Through reading and using the teacher's guide, teachers

9. Project Slide 8: Dinosaur Extinction Evidence Card A and continue the Science Seminar. As needed, remind students of their roles and responsibilities. Then, begin the discussion again. Ask, "What does this evidence tell us?"

> Possible prompt: You may wish to prompt students by considering each part of the evidence provided. Ask, "How large and how deep are the dimensions provided?" Ask students to compare the dimensions to known things. Discuss what a crater this size might mean. Discuss the significance of the date (65 mya).

10. Continue facilitating discussion. Although the aim is for students to run the conversation, if discussion has stalled, there are several ways you can facilitate.

> Possible prompt: Slide 9, Evidence Card B. You may want to prompt students by saying, "The volcanoes erupted over a short period of time. Does this change how you think about the volcanoes claim? Why?"

Fig. 13.1 Example of lesson plan element that supports teacher PCK

could develop their PCK of argumentation related to student difficulties. For example, in Lesson 2.10 of the *Rocks* unit, students are participating in a whole class oral argumentation activity called a Science Seminar, for which they have prepared by reading a variety of evidence sources. The seminar is focused on the evidence from rock layers and other sources that can support the claim that either the dinosaurs became extinct because of an asteroid hitting Earth or they became extinct because of volcanic eruptions. The teacher's role is to initially support the student-led discussion by providing prompts and then stepping aside to allow students to continue (see steps from the lesson plan in Fig. 13.1).

The prompts are constructed so that they provide a way for the teacher to offer support while still maintaining the role as facilitator and encouraging conversation and student discussion, rather than right answers. PCK for argumentation can be affected as the teacher takes up these kinds of prompts and sees how the students respond to them and engage with each other while analyzing evidence.

4. **Student-facing materials.** Although the purpose of the student-facing materials is obviously to support student learning, these materials also have the potential to support teacher learning. For example, in the curriculum, students are presented with graphics and text that highlight important aspects of argumentation, related to the student difficulties identified in the four conceptions. As one example, in *Space* 1.1, students view a projected image of a diagram that shows the relationship between a claim, evidence, and reasoning (see Fig. 13.2).

13.3.4   PCK Supports in Text-Based Educative Features

In addition to the basic step-by-step instructions for teaching a lesson, the curriculum included educative notes. These were text notes, appearing in a sidebar (see Fig. 13.3) that provided additional background, rationale, and guidance to teachers. Teachers could view the title of the note and then click on it to open up

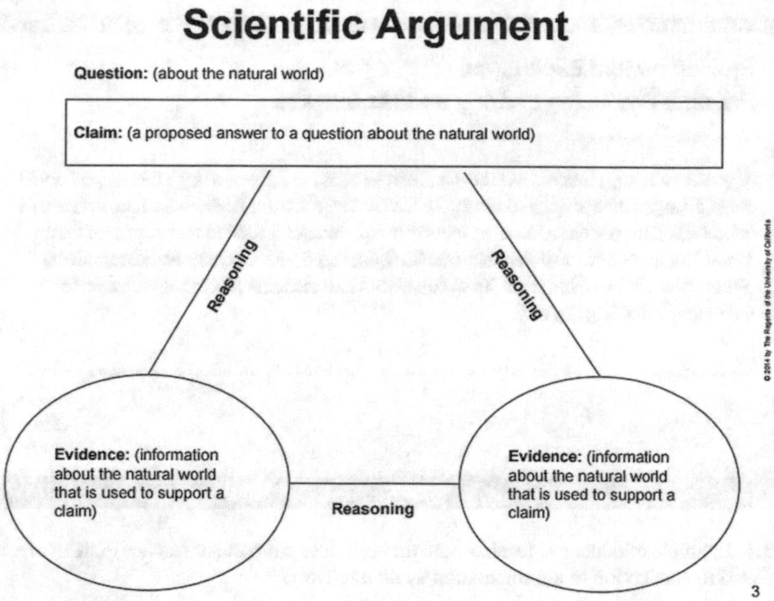

Scientific Argument

Question: (about the natural world)

Claim: (a proposed answer to a question about the natural world)

Reasoning

Reasoning

Evidence: (information about the natural world that is used to support a claim)

Reasoning

Evidence: (information about the natural world that is used to support a claim)

Fig. 13.2 Example of student-facing materials that also support teacher PCK for argumentation: a projected diagram representing the relationship between claims, evidence, and reasoning

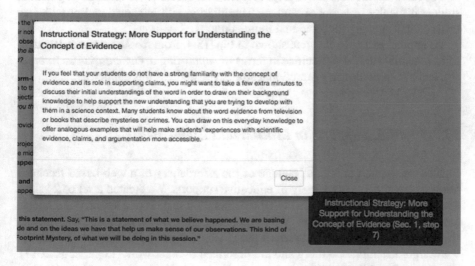

Fig. 13.3 Example of educative teacher note that provides support for teacher PCK of argumentation, related to conception of evidence

the note and read the full text. Unlike the step-by-step teacher's guide, these notes are not a mandatory component of teaching the lesson. Rather, they provide "meta-level" guidance to teachers. These notes are similar to educative elements that have been used in other curriculum programs (Cervetti et al. 2015; Davis and Krajcik 2005).

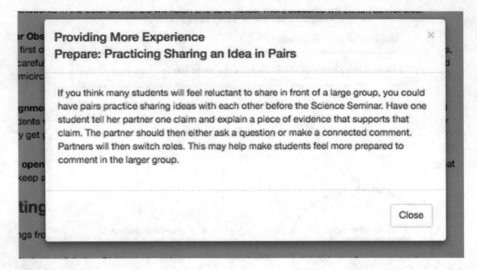

Fig. 13.4 Example of educative teacher note that provides support for teacher PCK of argumentation, related to conception of argumentation as interactive

For example, the note above (Fig. 13.3) in *Rocks* Lesson 1.2 identifies a potential student difficulty in terms of students' familiarity with evidence and provides an instructional strategy for providing more supports.

As another example, the note shown in Fig. 13.4, from *Rocks* Lesson 2.10, identifies potential student difficulties in terms of *supporting argumentation as interactive* and also provides an instructional strategy to address this challenge.

13.3.5 PCK Supports in Multimedia Elements

Finally, we wanted to take advantage of the affordances of a web-based teacher's guide to include a richer variety of educative supports. We created a set of 24 short educative videos, each embedded in the preparation section of a lesson at a strategic location within the curriculum. We also created slideshows, podcasts, and interactive reflective elements. Examples of the videos and other elements can be seen at www.argumentationtoolkit.org and are described in other work (Loper et al. 2017). The types of multimedia elements are summarized below.

Videos. Twenty-four short (2–6-min) videos were developed, focused around the identified student challenges of high-quality evidence, reasoning, competing claims, and student interaction. Videos fell into four categories: *rationale* videos that provided rationale for including argumentation in science class; *approach* videos that introduced the four conceptions and common student challenges associated with them; *activity* videos that introduced teachers to activities related to argumentation, included footage of real students engaging in the activities, and provided support to teachers for successfully enacting those activities; and *strategy* videos that elaborated on a specific instructional strategy for teachers to use to support students with challenging aspects of argumentation. Each video was embedded in the preparation section of a lesson in which that video was relevant. More detail about the development of the videos can be found in Loper et al. (2017).

Podcasts. To take advantage of the multimodality available in a web-based teacher's guide, we created a podcast for each unit that helped describe the goals and structure of the unit with respect to argumentation. Each podcast was between 1.5 and 2.5 min long. The idea was that an audio podcast provided teachers with flexibility in terms of when they listened; for example, teachers could listen while driving to work or while setting up their classroom.

Slideshows. We created four slideshows, each with text and images, that provided teachers with another means to access information related to argumentation. These were also embedded in the preparation section of a relevant lesson.

Interactive Reflective Elements. In order to make teachers' use of the multimedia educative elements a more interactive experience, we incorporated a number of interactive elements. Every embedded video was presented in conjunction with a reflection question, to prompt teachers to connect the video to their own teaching experience. In addition, a number of reflective self-assessment prompts were included in the curriculum. These were short survey-style questions that asked teachers to reflect on their own teaching practice related to argumentation. Depending on teachers' responses, different videos or other educative resources were recommended.

13.4   Methods

To address our research questions, we conducted a randomized experimental study with both a control group and pretest (Shadish et al. 2002). Teachers were randomly assigned to one of two conditions: half the teachers used educative curriculum materials that included only the text-based supports (i.e., PCK supports embedded in lesson plans and the educative teacher notes), and half the teachers used educative curriculum materials that included, in addition to the text-based supports, the multimedia elements. Teachers were asked to use the curriculum materials in whatever way they would normally use curriculum and completed a pre- and post-assessment of their PCK for argumentation.

13.4.1   Participants

A total of 90 middle-school science teachers from across the United States completed the study. Teachers were recruited via teacher networks and Listservs associated with the Lawrence Hall of Science. Teachers were excluded from the study if

Table 13.2 School background information ($n = 90$)		Text and multimedia ($n = 46$)	Text support only ($n = 44$)
Type of school			
	Public	43	38
	Private	2	2
	Charter	1	2
	Faith-based	0	1
	Other	0	1
School locale			
	City	16	14
	Suburb	10	13
	Town	6	6
	Rural	14	10
	Not sure	0	1

they had prior experience using curriculum materials associated with this project. The teachers came from public, private, and charter schools and represented a range in terms of years teaching science, educational background, and experience teaching scientific argumentation (see Tables 13.2 and 13.3). After being admitted to the study, teachers were randomly assigned to the text and multimedia support group or the text support only group. Random assignment was done in blocks at the school level, so if two or more teachers from the same school were participating, they were assigned to the same condition. After random assignment had been completed, a comparison was made between the two groups to make sure they were not significantly different in terms of type of school, school locale, years teaching science, and education level resulting in similar demographics (see Tables 13.2 and 13.3).

Each teacher was provided with a password-protected version of the Earth and Space Science curriculum site. The site included PDF copymasters of student sheets to copy and use. In addition, in order to support use of the curriculum, teachers were mailed kits containing most of the physical materials required to teach the lesson (e.g., rock samples).

13.4.2 Data Collection

All participants completed a pre- and post-survey that included 16 multiple-choice questions related to PCK for argumentation. An example of a multiple-choice item from the PCK measure is shown in Fig. 13.5 and illustrates how we used vignettes to target PCK of argumentation in action. The development of this measure is described in detail in McNeill et al. (2016a).

Table 13.3 Teacher background information ($n = 90$)

	Text and multimedia ($n = 46$)	Text support only ($n = 44$)
Years teaching science		
0–2 years	6	6
3–5 years	9	8
6–10 years	12	14
11–15 years	8	9
16+ years	11	7
Highest level of science education		
No response	4	1
Bachelor's degree	38	31
Master's degree	3	12
Doctorate degree	1	0
Highest level of education		
No response	1	1
Bachelor's degree + teaching certificate	26	14
Master's degree	18	29
Doctorate degree	1	0
Gender		
Male	6	9
Female	39	34
I do not wish to respond	1	1
Race/ethnicity		
White	40	32
Hispanic or Latino	2	0
Black or African American	0	1
Asian/Pacific Islander	1	2
Multiracial	2	7
Other	0	1

In addition to the PCK measure, we asked additional post-survey questions and collected back-end data on teacher use of the materials. The post-survey included several questions about how teachers used the online materials. In addition, we collected back-end data around teachers' use of the online curriculum guide including teachers' page views (for both groups) and video views (for the teachers in the text + multimedia group).

13.4.3 Analysis

To investigate Research Question 1, we used the data about the page views and the survey questions about curriculum use. For the quantitative items, we generated descriptive statistics. Open-ended survey responses were coded and categorized. We created graphical summaries of the results in order to compare across groups.

Mr. Cedillo's 7th grade science class is doing a unit, on force and motion. Near the middle of the unit his students are exploring friction by analyzing the data table from an investigation they conducted that answered the question: Which type of surface material will allow a toy car to have the greatest average speed? The students let a toy car go from the top of a ramp and timed how long it took to travel 1 m after reaching the bottom of the ramp, over four different surface materials: felt, top of lab table, sand paper, and ice (see image below).

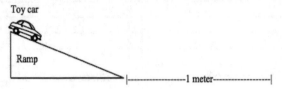

They then calculated the toy car's average speed by dividing the distance over the time. The table below shows the students' experimental results.

Surface material	Distance traveled (m)	Time (s)	Average speed (m/s)
Felt	1.0	2.4	0.42
Top of lab table	1.0	1.5	0.67
Sand paper	1.0	2.2	0.45
Ice	1.0	1.0	1.0

Ellen raises her hand in class and states the following argument: The car on the ice will always go the fastest. I've been in a car driving on ice, and I know a car can skid because ice is the smoothest surface. My dad has a really big truck and it doesn't slide as far, so maybe next time we should try this experiment with larger cars.

1. Mr. Cedillo should respond by saying:
    a. "Interesting point, Ellen. Does anyone have similar reasoning?"
    B. "Great connection. Can anyone suggest data to support this?"
    c. "Nice argument. What additional evidence could Ellen add?"
    d. "Well done. Does anyone else want to share their argument?"

Fig. 13.5 Example item from PCK for argumentation measure

For Research Questions 2 and 3, we first used a Rasch model to examine the validity of the PCK assessment (Andrich 1988). The variable map, which shows the distribution of item difficulties and person abilities, indicated that in general, the items covered the range of respondents' ability reasonably well. The instrument appeared to have high item separation (>3, item reliability >0.9), indicating that the person sample and item difficulty range are large enough to establish the item difficulty hierarchy. Based on the goodness-of fit statistics (standardized statistics and mean square statistics), none of the items appeared to be problematic. Moreover, Mantel-Haenszel tests conducted to examine differential item functioning (DIF) by respondents' gender indicated no significant DIF statistics for any item. However, the person separation was low (<2, person reliability <0.8), which implies that the instrument may not be sensitive enough to distinguish between individuals who are high and low on the PCK of argumentation construct. Taken together, these results suggest that, while this instrument shows promise for measuring PCK of argumentation, there are too few easy and difficult items included on the instrument in its current form. Consequently, we continued with our analysis since none of the items were problematic; however, we were aware that the measure was limited in its ability to measure growth or change, which were the focus of Research Questions 2 and 3.

For Research Question 2, examining whether the educative curriculum impacted all 90 teachers' PCK of argumentation, we used a paired samples t-test to see if there were significant changes from the pretest to the posttest. We were interested in whether there was teacher growth after enacting the curriculum.

For Research Question 3, examining the impact of the multimedia educative curriculum on teachers' PCK of argumentation, we used multiple linear regression. Specifically, teachers' posttest PCK scores were regressed on their pretest PCK scores, an indicator of their group membership, and a series of background characteristics and back-end use. A power analysis was conducted using G*Power to estimate the minimum effect size detectable with a linear regression model given a fixed sample size, power, and significance level. With 90 teachers, power = 0.80, and $\alpha = 0.05$, a linear regression model with a maximum of four predictors will allow us to detect a medium effect size of $f^2 = 0.15$. Given the relatively limited sample size and statistical power, the models were formulated sequentially to include three blocks of variables: demographics, condition, and back-end use. For each outcome, a parsimonious final model that included only group membership and any statistically significant predictors from previous blocks was formulated.

13.5   Results

13.5.1   Research Question 1: Use of the Materials

As mentioned above, teachers were told to use the curriculum in whatever way they would normally use curriculum materials. To get a sense of how extensively teachers used the web-based curriculum materials, we counted the number of page views each teacher had for the website. The site had a total of 72 pages (one for each of 62 lessons, as well as 10 additional pages with overview information). Page views varied widely, with an average of around 300 views for both groups, but a range from as low as 14 to as high as 799. Figure 13.6 shows the page views for both the text-only and text + multimedia groups below.

Figure 13.7 shows, for the teachers who received the multimedia supports, how often they played educative videos within the digital curriculum. This shows a wide range of use from six teachers who played no videos to four teachers who played over 21 videos.

Teachers were also asked about how they used the curriculum. As shown in Fig. 13.8, the most common way that teachers used the curriculum was to use a combination of using the website and printing or creating materials. This suggests

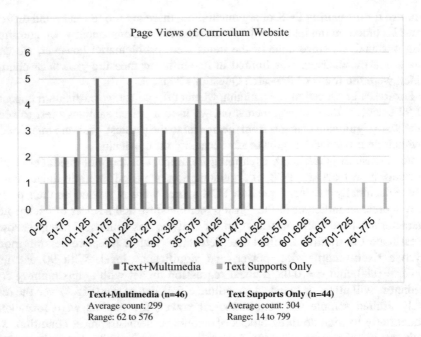

Fig. 13.6 Page views across groups

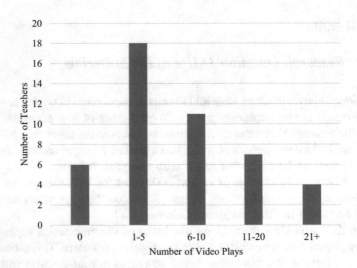

Fig. 13.7 Number of video plays ($n = 46$)

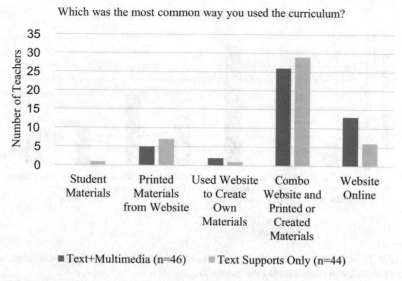

Fig. 13.8 Ways teachers used the curriculum

that while the teachers in the text + multimedia group had access to the videos and other multimedia resources, the way they used the curriculum (e.g., printed versus website only) may have impacted how often they used those resources. In contrast, most of the text-based supports would have still been available in the printed form.

Teachers' open-ended responses to questions about why they used the curriculum in the above ways were categorized and are shown in Fig. 13.9 below. This suggests that some of the reasons teachers did not use just the web-based materials were because they wanted to make adaptations, found limitations with it for preparation, had challenges with technology, or simply preferred paper.

13.5.2 Research Question 2: Impact of the Educative Curriculum Materials

Pre- and posttest PCK scores were created by calculating the sum for each teacher across the 16 PCK items. Table 13.4 presents the mean pre- and posttest scores, along with the results of the paired samples t-test. The results indicate that teachers' posttest PCK scores were statistically significantly higher than teachers' pretest PCK scores. Considering the limitations of the measure to detect growth that were revealed by the Rasch analysis, this is particularly promising. This suggests that enacting educative curriculum materials can support teachers' PCK of argumentation.

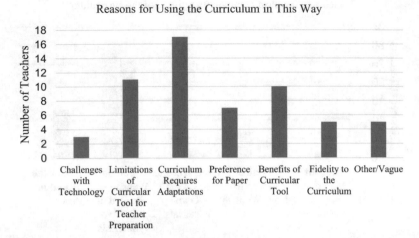

Fig. 13.9 Reasons why teachers used the web-based curriculum in a particular way (note, teachers' responses could receive more than one code)

Table 13.4 Descriptive statistics and paired samples *t*-test

		Pretest		Posttest		Paired samples *t*-test	
	N	Mean (se)	SD	Mean (se)	SD	t	Sig.
PCK scores	90	8.47 (0.25)	2.42	9.00 (0.24)	2.31	−2.31	0.023*

*Significant for $p < 0.05$

13.5.3 Research Question 3: Impact of the Additional Multimedia Elements

Finally, we ran a regression model to examine the effect of the text + multimedia elements condition on teachers' PCK of argumentation. Table 13.5 includes the results from this analysis. The first model includes teacher demographic information to control for any variation that did exist between the two groups. Here we do see that some of these variables significantly impacted teachers' post PCK of argumentation scores. For example, teachers with higher pretest scores and teachers with fewer years teaching experience had higher posttest scores. In addition, white teachers ($n = 72$) had higher post PCK of argumentation scores compared to non-white teachers ($n = 18$). These three variables explained 32.4% of the variation in teachers' post PCK of argumentation score.

Unfortunately, none of the variables in the next two models significantly impacted teachers' posttest scores. Specifically, the condition, number of page views, and numbers of videos played did not significantly impact teachers' PCK of argumentation as evaluated by our assessment measure. There are a number of possible reasons for the lack of significance of these variables. First of all, as the

Table 13.5 Regression model for teachers' PCK of argumentation

	Model 1 Demographics	Model 2 Condition	Model 3 Back-end use
	Coeff (s.e.)	Coeff (s.e.)	Coeff (s.e.)
Intercept	5.9 (0.91)***	5.63 (0.9)***	5.59 (0.97)***
Demographics			
Pretest PCK	0.48 (0.09)***	0.49 (0.08)***	0.48 (0.09)***
Years teaching	−0.34 (0.16)*	−0.33 (0.15)*	−0.33 (0.16)*
Degree	0.15 (0.44)		
Race	−1.29 (0.54)*	−1.19 (0.53)*	−1.17 (0.56)*
Gender	−0.19 (0.38)		
Condition			
Group		0.41 (0.41)	0.44 (0.5)
Back-end use			
Total page views[a]			<0.001 (<0.01)
Total video views			<0.01 (0.04)
Adjusted R^2	0.324	0.355	0.339

***Significant at $p < 0.001$, *significant at $p < 0.05$
[a]Variable rescaled to represent a change of ten units

Rasch analysis suggested, our measure of PCK of argumentation is not very sensitive to growth. Consequently, a different or revised assessment might detect differences between the two groups. In addition, the curriculum use data suggests the teachers used the curriculum in many ways, including relying on paper versions of the materials, which would not enable them to watch the videos. This suggests the importance of addressing some of the limitations of the digital curriculum to support greater teacher use.

13.6   Discussion

In this work, we found the notion of PCK of argumentation to be a helpful framework for the design of educative curriculum materials, including both text-based and multimedia features, and we found preliminary evidence suggesting that these educative curriculum materials supported development of teachers' PCK, even in the absence of professional development. We also documented teachers' variable use of web-based educative curriculum materials. These findings have implications for future efforts to design curriculum materials and support teachers' PCK development. In addition, we identified challenges with the assessment of PCK for argumentation that can inform future research efforts and measure development.

Educative curriculum materials can support PCK. Efforts to support teachers' development of PCK have primarily focused on pre-service teachers (Schneider and Plasman 2011) or on in-service teachers through professional development (e.g., McNeill and Knight 2013; Osborne et al. 2004; Simon et al. 2006). Educative curriculum materials, because they can be more easily scaled up to be used by large numbers of teachers in the context of their daily work, could efficiently reach large numbers of in-service teachers. Previous work suggests that educative curriculum materials hold promise for supporting teacher learning (Arias et al. 2016b; Davis and Krajcik 2005; Gess-Newsome et al. 2011) and teacher practice (Arias et al. 2016a; Cervetti et al. 2015). However, there has been little empirical work examining the impact of educative curriculum materials on teachers' PCK. As discussed previously, we emphasize a dynamic view of PCK that focuses on knowledge-in-use (Alonzo and Kim 2016; Cochran et al. 1993). Since educative curriculum materials are by their nature situated in use and linked to classroom practice, they may be well-suited to support teachers' development of dynamic PCK. The findings presented here suggest that educative curriculum materials can support teachers' PCK for argumentation. Specifically, we found that after enacting the curriculum, teachers were better able to evaluate the strengths and weaknesses of students' arguments and suggest appropriate instructional strategies to meet students' needs. Although the educative curriculum supported teacher learning, we did not detect a significant impact of the additional multimedia elements.

Teachers' variable use of web-based educative curriculum materials must be taken into account. Prior work related to educative curriculum materials and teacher learning has focused on print-based materials (Arias et al. 2016b; Cervetti et al. 2015). The rapid shift in curriculum materials from paper to digital formats opens up questions about how teachers' learning will shift with these new technologies (Drake, Land, & Tyminski, 2014). Thus, a better understanding of how teachers use web-based curriculum materials is needed in order to maximize the materials' potential impact on teacher PCK. As shown in the results above, teachers varied widely in how, and how much, they used the web-based curriculum materials. Overall, the text-only and text + multimedia groups were similar, but there are some interesting differences. The teachers in the text + multimedia group were more likely to report that they primarily used the curriculum materials online (as opposed to printing them out). They were also more likely to refer to the benefits of the web-based curriculum than the teachers using the text-only materials. Finally, they were less likely than the teachers using the text-only materials to express a preference for paper. These differences are not statistically significant but may suggest the possibility that the addition of the multimedia elements helped teachers to see the value of using a web-based, digital teacher's guide.

However, the most common way to use the materials across both groups was a combination of the website and printed or created materials. This suggests that in their current form, there were limitations to the design of the digital materials. Teachers reported a need for better ways to adapt the materials and use them more flexibly for preparation. Increasing the flexibility and interactivity of the curriculum materials could address some of these concerns, which would support greater use.

In addition, greater interactivity of online learning environments can support deep cognitive processing in the learner (Moreno and Mayer 2007) and provide more control to learners which supports self-regulation (Song and Hill 2007). Thus, improvements to the usefulness, flexibility, and interactivity of web-based educative curriculum materials could potentially increase their impact on PCK.

13.7 Limitations and Future Directions

The study had a number of limitations. For example, because we did not have a pure control—teachers who did not use any version of the curriculum—we cannot be certain that teachers' pre-/post growth was not a result of learning that occurred unrelated to the curriculum use. Future efforts could include such a control to determine the effect of using educative curriculum related to business-as-usual curriculum. In addition, we are interested in exploring whether the benefit of the multimedia educative features would be enhanced if they were made more customizable and interactive. Web-based educative curriculum materials offer potential for new approaches to teacher learning that are worthy of further exploration (Loper et al. 2017).

Similar to other researchers (e.g., Park and Oliver 2008), we found that PCK is challenging to assess. Although there were no problems with any of our individual items, overall our assessment is not sensitive enough to distinguish between high and low PCK, which is important for evaluating growth. While this instrument shows promise for measuring PCK of argumentation, we need to add more easy and difficult items to the instrument to the extremes of the ability range to increase the person separation and increase the sensitivity of the measure for discriminating among respondents.

13.8 Implications

The finding that educative curriculum materials can impact teachers' PCK is an important one because ECMs can be efficiently integrated into teachers' daily teaching practice. Thus, if well-designed educative curriculum materials can have even a modest impact on teachers' PCK, this could be a practical and cost-effective way to support teacher learning. However, we must also consider the implications of our findings of wide variability in teachers' use and access of the materials. Teachers' preferences and patterns of use should be taken into account in order to design materials that will meet teachers' needs and thus maximize the likelihood that teachers will access and benefit from educative elements in the curriculum.

References

Abell, S. K. (2008). Twenty years later: Does pedagogical content knowledge remain a useful idea? *International Journal of Science Education, 30*(10), 1405–1416.

Alonzo, A. C., & Kim, J. (2016). Declarative and dynamic pedagogical content knowledge as elicited through two video-based interview methods. *Journal of Research in Science Teaching, 53*(8), 1259–1286.

Alozie, N. M., Moje, E. B., & Krajcik, J. S. (2010). An analysis of the supports and constraints for scientific discussion in high school project-based science. *Science Education, 94*, 395–427.

Andrich, D. (1988). *Rasch models for measurement*. Newbury Park, CA: Sage Publications Inc.

Arias, A., Bismack, A., Davis, E. A., & Palincsar, A. S. (2016a). Interacting with a suite of educative features: Elementary science teachers' use of educative curriculum materials. *Journal of Research in Science Teaching, 53*(3), 442–449.

Arias, A. M., Davis, E. A., Marino, J. C., Kademian, S. M., & Palincsar, A. S. (2016b). Teachers' use of educative curriculum materials to engage students in science practices. *International Journal of Science Education, 38*(9), 1504–1526.

Avraamidou, L., & Zembal-Saul, C. (2005). Giving priority to evidence in science teaching: A first-year elementary teacher's specialized practices and knowledge. *Journal of Research in Science Teaching, 42*(9), 965–986.

Ball, D. L., & Cohen, D. K. (1996). Reform by the book: What is—or might be—the role of curriculum materials in teacher learning and instructional reform? *Educational Researcher, 25*, 6–8, 14.

Berland, L. K., & Reiser, B. J. (2011). Classroom communities' adaptations of the practice of scientific argumentation. *Science Education, 95*, 191–216.

Beyer, C., Delgado, C., Davis, E., & Krajcik, J. (2009). Investigating teacher learning supports in high school biology curricular programs to inform the design of educative curriculum materials. *Journal of Research in Science Teaching, 46*(9), 977–998.

Brown, M. (2009). The teacher-tool relationship: Theorizing the design and use of curriculum materials. In J. Remillard, G. Lloyd, & B. Herbel-Eisenmann (Eds.), *Teachers' use of mathematics curriculum materials: Research perspectives on relationships between teachers and curriculum*. New York, NY: Routledge.

Cavagnetto, A. R. (2010). Argument to foster scientific literacy: A review of argument interventions in K–12 science contexts. *Review of Educational Research, 80*(3), 336–371.

Cervetti, G. N., Kulikowich, J. M., & Bravo, M. A. (2015). The effects of educative curriculum materials on teachers' use of instructional strategies for English language learners in science and on student learning. *Contemporary Educational Psychology, 40*, 86–98.

Cochran, K. F., DeRuiter, J. A., & King, R. A. (1993). Pedagogical content knowing: An integrative model for teacher preparation. *Journal of Teacher Education, 44*(4), 263–272.

Davis, E. A., & Krajcik, J. S. (2005). Designing educative curriculum materials to promote teacher learning. *Educational Researcher, 34*(3), 3–14.

Davis, E. A., Palincsar, A. S., Arias, A. M., Bismack, A. S., Marulis, L. M., & Iwashyna, S. K. (2014). Designing educative curriculum materials: A theoretically and empirically driven process. *Harvard Educational Review, 84*(1), 24–52.

Davis, E. A., Janssen, F. J. J. M., & Van Driel, J. H. (2016). Teachers and science curriculum materials: Where we are and where we need to go. *Studies in Science Education, 52*(2), 127–160.

Drake, C., Land, T. J., & Tyminski, A. M. (2014). Using educative curriculum materials to support the development of prospective teachers' knowledge. *Educational Researcher, 43*(3), 154–162.

Gess-Newsome, J., Cardenas, S., Austin, B. A., Carlson, J., Gardner, A. L., & Wilson, C. D. (2011). *Impact of educative materials and transformative professional development on teachers PCK, practice, and student achievement*. Paper presented at the NARST annual meeting, Orlando, FL.

Hume, A., & Berry, A. (2011). Constructing CoRes—a strategy for building PCK in pre-service science teacher education. *Research in Science Education, 41*(3), 341–355.

Jiménez-Aleixandre, M. P., & Erduran, S. (2008). Argumentation in science education: An overview. In S. Erduran & M. P. Jimenez-Aleixandre (Eds.), *Argumentation in science education: Perspectives from classroom-based research* (pp. 3–28). Dordrecht, The Netherlands: Springer.

Kind, V. (2009). Pedagogical content knowledge in science education: Perspectives and potential for progress. *Studies in Science Education, 45*(2), 169–204.

Loper, S., McNeill, K. L., & González-Howard, M. (2017). Multimedia educative curriculum materials (MECMs): Teachers' use of MECMs to support argumentation. *Journal of Science Teacher Education, 28*(1), 36–56.

Magnusson, S., Krajcik, J., & Borko, H. (1999). Nature, sources, and development of pedagogical content knowledge for science teaching. In J. Gess-Newsome & N. G. Lederman (Eds.), *Examining pedagogical content knowledge* (pp. 95–132). Dordrecht, The Netherlands: Springer.

McNeill, K. L., & Knight, A. M. (2013). Teachers' pedagogical content knowledge of scientific argumentation: The impact of professional development on k-12 teachers. *Science Education, 97*(6), 936–972.

McNeill, K. L., Lizotte, D. J., Krajcik, J., & Marx, R. W. (2006). Supporting students' construction of scientific explanations by fading scaffolds in instructional materials. *The Journal of the Learning Sciences, 15*(2), 153–191.

McNeill, K. L., González-Howard, M., Katsh-Singer, R., & Loper, S. (2016a). Pedagogical content knowledge of argumentation: Using classroom contexts to assess high quality PCK rather than pseudoargumentation. *Journal of Research in Science Teaching, 53*(2), 261–290.

McNeill, K. L., Marco-Bujosa, L. M., González-Howard, M., & Loper, S. (2016b, April). *Curriculum implementation for scientific argumentation: Fidelity to procedure versus fidelity to goal.* Paper presented at the annual meeting of the National Association for Research in Science Teaching, Baltimore, MD.

McNeill, K. L., González-Howard, M., Katsh-Singer, R., & Loper, S. (2017). Moving beyond pseudoargumentation: Teachers' enactments of an educative science curriculum focused on argumentation. *Science Education, 101*(3), 426–457.

Moreno, R., & Mayer, R. (2007). Interactive multimodal learning environments. *Educational Psychology Review, 19*(3), 309–326.

National Research Council. (2012). *A framework for K-12 science education: Practices, crosscutting concepts, and core ideas.* Committee on a Conceptual Framework for New K-12 Science Education Standards. Board on Science Education, Division of Behavioral and Social Sciences and Education. Washington, DC: The National Academies Press.

NGSS Lead States. (2013). *Next generation science standards: For states, by states* (Appendix F). Washington, DC: The National Academies Press.

Osana, H. P., & Seymour, J. R. (2004). Critical thinking in pre-service teachers: A rubric for evaluating argumentation and statistical reasoning. *Educational Research and Evaluation, 10*(4–6), 473–498.

Osborne, J. (2010). Arguing to learn in science: The role of collaborative, critical discourse. *Science, 328*, 463–466.

Osborne, J., Erduran, S., & Simon, S. (2004). Enhancing the quality of argumentation in school science. *Journal of Research in Science Teaching, 41*(10), 994–1020.

Park, S., & Oliver, S. (2008). Revisiting the conceptualisation of pedagogical content knowledge (PCK): PCK as a conceptual tool to understand teachers as professionals. *Research in Science Education, 38*, 261–284.

Pruitt, S. L. (2014). The next generation science standards: The features and challenges. *Journal of Science Teacher Education, 25*(2), 145–156.

Remillard, J. T. (2005). Examining key concepts in research on teachers' use of mathematics curricula. *Review of Educational Research, 75*(2), 211–246.

Roth, K. J., Garnier, H. E., Chen, C., Lemmens, M., Schwille, K., & Wickler, N. I. (2011). Videobased lesson analysis: Effective science PD for teacher and student learning. *Journal of Research in Science Teaching, 48*(2), 117–148.

Sadler, T. D. (2006). Promoting discourse and argumentation in science teacher education. *Journal of Science Teacher Education, 17*, 323–346.

Sampson, V., & Blanchard, M. (2012). Science teachers and scientific argumentation: Trends in views and practice. *Journal of Research in Science Teaching, 49*, 1122–1148.

Sampson, V., & Clark, D. (2008). Assessment of the ways students generate arguments in science education: Current perspectives and recommendations for future directions. *Science Education, 92*, 447–472.

Schneider, R. M., & Plasman, K. (2011). Science teacher learning progressions: A review of science teachers' pedagogical content knowledge development. *Review of Educational Research, 81*(4), 530–565.

Settlage, J. (2013). On acknowledging PCK's shortcomings. *Journal of Science Teacher Education, 24*, 1–12.

Shadish, W. R., Cook, T. D., & Campbell, D. T. (2002). *Experimental and quasi-experimental designs for generalized causal inference*. Boston, MA: Houghton Mifflin Company.

Sherin, M. G. (2003). Using video clubs to support conversations among teachers and researchers. *Action in Teacher Education, 24*(4), 33–45.

Shulman, L. S. (1986). Those who understand: Knowledge growth in teaching. *Educational Researcher, 15*(2), 4–14.

Simon, S., Erduran, S., & Osborne, J. (2006). Learning to teach argumentation: Research and development in the science classroom. *International Journal of Science Education, 28*(2–3), 235–260.

Song, L., & Hill, J. R. (2007). A conceptual model for understanding self-directed learning in online environments. *Journal of Interactive Online Learning, 6*(1), 27–42.

Toulmin, S. (1958). *The uses of argument*. Cambridge, UK: Cambridge University Press.

van den Berg, E., Wallace, J., & Pedretti, E. (2008). Multimedia cases, teacher education and teacher learning. In J. Voogt & G. Knezek (Eds.), *International handbook of information technology in primary and secondary education* (pp. 475–487). New York, NY: Springer.

Watters, J. J., & Diezmann, C. M. (2007). Multimedia resources to bridge the praxis gap: Modeling practice in elementary science education. *Journal of Science Teacher Education, 18*(3), 349–375.

Wilson, S. M. (2013). Professional development for science teachers. *Science, 340*(6130), 310–313.

Zembal-Saul, C. (2009). Learning to teach elementary school science as argument. *Science Education, 93*, 687–719.

Zembal-Saul, C., Munford, D., Crawford, B., Friedrichsen, P., & Land, S. (2002). Scaffolding preservice science teachers' evidence-based arguments during an investigation of natural selection. *Research in Science Education, 32*(4), 437–465.

Zohar, A. (2008). Science teacher education and professional development in argumentation. In S. Erduran & M. P. Jimenez-Aleixandre (Eds.), *Argumentation in science education: Perspectives from classroom-based research* (pp. 245–268). Dordrecht, The Netherlands: Springer.

Chapter 14
Teacher Education for Maker Education: Helping Teachers Develop Appropriate PCK for Engaging Children in Educative Making

Danielle B. Harlow, Alexandria K. Hansen, Jasmine K. McBeath, and Anne E. Leak

Abstract Despite the potential of the maker movement to influence how we teach students in school, thus far, most research on maker activities have taken place in informal spaces, such as museums and after-school programs, which are inaccessible to some populations. To ensure maker education reaches *all* students, it must find its place at school. However, classroom-based maker activities have different constraints and may require teachers to hold different types of knowledge. We drew from the body of research on maker education to create a course that prepared pre-service elementary school teachers to implement activities that were consistent with the maker ethos and met state and district standards. As a course assignment, the teacher candidates designed and hosted a School Maker Faire for elementary school children, providing an opportunity for local children to participate in maker activities and for pre-service elementary school teachers to design, facilitate, and reflect on maker education as a method of teaching science. In this paper, we delineate the constituent parts of maker pedagogical content knowledge and describe how pre-service teachers developed the appropriate knowledge for integrating maker education activities into their classroom curriculum. We propose that the knowledge teachers need to facilitate and assess student learning through maker education is more complex than either science pedagogical content knowledge or engineering pedagogical content knowledge.

Keywords Maker PCK · Teacher education · Maker education · NGSS · Educative making · Materials knowledge · Elementary science

D. B. Harlow (✉) · A. K. Hansen · J. K. McBeath
Department of Education, UC-Santa Barbara, Santa Barbara, CA, USA

A. E. Leak
School of Physics and Astronomy, Rochester Institute of Technology, Rochester, NY, USA

© Springer International Publishing AG, part of Springer Nature 2018
S. M. Uzzo et al. (eds.), *Pedagogical Content Knowledge in STEM*,
Advances in STEM Education, https://doi.org/10.1007/978-3-319-97475-0_14

14.1 Introduction

The engineering workforce does not represent the diversity of the US population. Women represent only 15% of the engineering workforce and 12% of physicists and astronomers. Hispanic Americans make up 16% of the workforce but only 6.6% of the engineering and 4.8% of the physical sciences workforce (NSF 2015). To change these statistics, we focus on elementary school education.

Most research and programs designed to increase diversity in STEM (Science, Technology, Engineering, and Mathematics) focus on high school students (Maltese and Tai 2011), yet this is too late. Research has consistently shown the importance of early interest (Tai et al. 2006; Maltese and Tai 2011; Simpkins et al. 2006). Despite clear evidence that early experiences are critical to encouraging students to pursue engineering careers, many students receive little science or engineering education prior to middle school (Dorph et al. 2011).

Transitioning to the Next Generation Science Standards [NGSS] (NGSS Lead States 2013), with disciplinary core ideas such as Engineering, Technology, and Applications of Science, will increase children's exposure to engineering. However, these changes will be challenging for elementary teachers to implement without support (Quinn and Bell 2013; Wardrip and Brahms 2016), especially in ways that will broaden diverse students' participation in these disciplines. New models of pedagogy and teacher development are needed.

Maker education has the potential to influence such a change in pedagogy. We consider maker education to be activities that engage students in "making" for the purposes of learning. Making includes "activities focused on designing, building, modifying, and/or repurposing material objects, for playful or useful ends, oriented toward making a 'product' of some sort that can be used, interacted with, or demonstrated" (Martin 2015, p. 31). Such activities are often "personal, local, and relevant" (Maltese and Tai 2011, p. 900) which is also consistent with culturally relevant pedagogy (Gay 2010) and recommendations for increasing the number of underrepresented students in STEM.

Bevan (2017) identified three areas of maker education: entrepreneurship, STEM workforce skills, and educative making. Here, we are concerned only with the last of these categories, educative making, "a pedagogical approach to engaging students in design-build activities that allow them to explore ideas, develop skills and understanding within particular (and often interdisciplinary) disciplines, and build a wide range of learning dispositions and capacities" (p. 6). We narrow our focus even further to consider only educative making designed to provide opportunities for elementary school students to develop knowledge aligned with the NGSS.

Making and engineering, while not identical, are related (Martinez and Stager 2013), and maker activities can develop engineering knowledge. Further, we know that learning through making appeals to a wide diversity of students (Blikstein et al. 2014; Petrich et al. 2013; Vossoughi et al. 2016), an important consideration since our goal is to increase participation of underrepresented populations. Making also provides opportunities for students to learn skills such as ingenuity, problem-solving, creativity, and teamwork, which are essential for success in STEM careers (Clough 2004; Mulvey and Pold 2015).

The movement to include maker activities in educational contexts is ripe for stimulating systematic change to reimagine education (Peppler and Bender 2013). However, many elementary teachers lack prior experience with engineering or making. As such, to enable effective change, teachers need support developing the appropriate expertise to facilitate such learning experiences. As Quinn and Bell (2013) noted, "Teachers who have never experienced these practices in their own science education will have difficulty implementing them in their classrooms" (p. 26).

Over the past several years, we have used the research on teacher education and maker education to iteratively design and research a series of activities with the goal of helping pre-service elementary school teachers develop the knowledge necessary to design, facilitate, and assess engineering and educative making activities. The iterative design and research on this course have helped to inform our understanding of the knowledge that teachers need to develop.

14.2   Teacher Education to Support Development of Maker PCK

The course we use to focus our discussion is an elementary science methods course which is part of a 13-month masters and teacher credential program. The science method course is held during the second half of the program and meets ten times for 3 h each meeting. The course is designed to provide pre-service elementary teachers with opportunities to learn about teaching and learning science and engineering in ways that support children's learning of the core ideas, practices, and crosscutting concepts included in the NGSS. The course design was guided by constructivism (Fosnot and Perry 1996; Fosnot 2013) and constructionism (Papert and Harel 1991). That is, we regularly provided opportunities for pre-service teachers to grapple with and make sense of science and engineering phenomena through active making and learning in stem-rich contexts.

We began integrating engineering into the course with the introduction of the Framework for Science Education (NRC 2012). In 2012, we introduced engineering to that year's cohort of pre-service teachers. With the cohort of the following year, we started discussing engineering and specifically how it fit into the NGSS. Over time, educative making became more central, as it became increasingly clear that engineering was the area of the NGSS that our pre-service teachers were least comfortable with and that educative making held potential as a tool for effective instruction. Starting with the class of 2016, making became a unifying theme for the course with a culminating School Maker Faire. It is these most recent years (courses offered in the Spring of 2016 and 2017) that we focus on here.

In designing the course, we found it valuable for the pre-service teachers to (1) experience educative making activities as learners, (2) consider the ways that informal educators think about facilitating learning and reflect on how these can be related to formal education, (3) examine resources and activities designed by informal science educators and match to NGSS and CCSS standards, (4) facilitate a maker activity with children, and (5) design and assess learning related to an educative making activity.

The learning experiences for the course were specifically designed to support pre-service elementary teachers in their efforts to design, facilitate, and assess a maker activity that engaged their students in science or engineering content. The culminating assignment for the pre-service teachers was to facilitate their activity at a School Maker Faire, hosted on the university's campus. While local teachers and administrators from the community attended the event to learn about educative making, the primary invited audience was elementary school students enrolled in the student teaching placement classrooms of the pre-service teachers (See Harlow and Hansen 2018, for a more complete description of the event).

The pre-service teachers worked in small groups of 2–4 to iteratively design an activity that they facilitated both in their classroom practicum placement and at the School Maker Faire. They were also required to develop an assessment activity and rubric to accompany the maker activity, which was specifically designed to elicit student ideas and measure development of NGSS-aligned content. The pre-service teachers could assess student learning at the School Maker Faire or in their practicum classroom placement. This assignment was designed as an explicit way for the pre-service teachers to link the activity they designed to the goals and constraints of the schools.

Over this 6-year period, starting from when we first began introducing engineering in 2012, we have implemented activities and assessed the outcomes through reflective assignments and observations of the pre-service teachers. As we identified gaps in the pre-service teachers' knowledge, and tracked their instructional struggles and successes, we developed a framework for the knowledge required to successfully implement educative maker activities aligned with the NGSS learning goals. In Sect. 14.3, we describe the model of pedagogical content knowledge required for educative making in science. Then in Sect. 14.4, we describe how we used this model to inform the specific educative making activities in the course.

14.3   Maker Pedagogical Content Knowledge (MPCK): A Model

The knowledge that teachers need in order to design, facilitate, and assess educative making is multifaceted. We drew on literature from science education, engineering education, and informal science education to develop experiences for pre-service teachers to develop the appropriate knowledge to integrate educative making activities into their curriculum. In this section, we describe what we considered the constituent parts of maker pedagogical content knowledge.

The two domains of teacher knowledge that are often emphasized are content knowledge (a focus on subject matter) and pedagogical knowledge (a focus on practical implementation in the classroom). Shulman (1987) advocated for combining these domains into a new model called pedagogical content knowledge, which represents a "blending of content and pedagogy into an understanding of how particular topics, problems, or issues are organized, represented, and adapted to the diverse interests and abilities of learners" (p. 8). Koehler and Mishra (2008) adapted Schulman's model to include a focus on technology in the digital age, proposing that there were three primary areas of knowledge that teachers needed to develop to effectively and flexibly use technology in their classroom: content knowledge, pedagogical knowledge, and technology knowledge. They called their model *technological pedagogical content knowledge (TPCK)*.

We consider the knowledge for educative making as a special case of TPCK. We call this Maker Pedagogical Content Knowledge (MPCK), and it differs from TPCK in all three areas of knowledge. The content knowledge required includes both science content knowledge and engineering content knowledge. The pedagogical knowledge for educative making also differs from conventional classroom practices. Educative making demands "new modes of classroom work, new material needs, new ways of guiding learning, and new demands on teachers' capacity to support learners" (Wardrip and Brahms 2016, p. 98). Finally, while TPCK primarily considers technology in the hands of the teacher (e.g., smart boards), in educative making, we are most concerned with the technology that is *in the hands of the learner*. For MPCK, materials knowledge extends beyond traditional classroom tools to include new digital fabrication tools such as 3D printers and laser cutters as well as technology such as circuits, robotics, and coding.

Below we briefly describe the contents of each of the circles in Fig. 14.1. However, discussing any one area in isolation of the others is not only difficult but likely unproductive. So, while we focus on one area at a time, the description of

Fig. 14.1 Components of Maker Pedagogical Knowledge

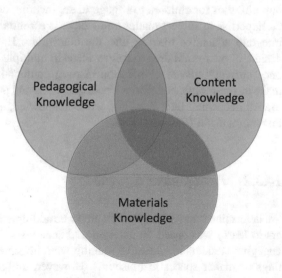

each is influenced by the other two. We then discuss how we helped pre-service teachers develop the beginnings of this knowledge base.

14.3.1 Content Knowledge

Educative making has the potential to encourage interest in and develop expertise in a wide variety of content areas including math (Garneli et al. 2013), art (Peppler 2016), writing (Cantrill and Oh 2016), and computing (Papert 1980). We, as science educators, focus on the potential educative making has for science education and, as such, consider only science content knowledge in this section. In the United States, science education is currently guided by the Next Generation Science Standards (NGSS Lead States 2013). The NGSS includes three interwoven stands: (1) Science and Engineering Practices, (2) Crosscutting Concepts, and (3) Disciplinary Core Ideas. These three types of goals are combined into performance expectations. Nearly all of the science disciplinary core ideas can be taught through activities that involve educative making or a combination of maker and non-maker education activities.

In addition to the content associated with the sciences, all grade levels include expectations for learning about engineering design. Engineering design is the process by which engineers develop innovative solutions to problems. The process involves understanding the problem, generating ideas, selecting an idea based on multiple constraints, and improving the idea. The NGSS breaks engineering design into three stages: (1) Defining and Delimiting an Engineering Problem, (2) Developing Possible Solutions, and (3) Optimizing the Design Solution. The specific expectations around these big ideas begin with simple expectations and become more complex over children's time in school. Maker activities can provide rich opportunities for children to engage in engineering design.

Expertise in subject matter knowledge is essential for any teacher attempting to integrate educative making into the classroom. Educative making activities, by design, allow for children to follow ideas in multiple directions, creating more than one "correct" answer or solution. Strong content knowledge allows teachers to make in-the-moment decisions about questions and prompts that will guide students toward deepening their thinking, while acknowledging that pathways students follow might differ (Sawyer 2004).

14.3.2 Pedagogical Knowledge

We determined that, while the *content* to be addressed (and thus necessary for teachers to learn) was based on national and state standards, the *pedagogical* tools for engaging students in educative making were informed by research in informal settings or maker spaces in museums. However, unlike museums, teachers are held accountable to national and state standards and must assess and document student learning along these dimensions. Thus, in this section, we talk about the design and facilitation strategies derived from work in museums and then discuss assessment strategies relevant to educative making for teachers.

Two museums that have well-documented descriptions of pedagogical strategies include the MAKESHOP within the Children's Museum of Pittsburgh (Wardrip and Brahms 2015) and the Tinkering Studio within the Exploratorium in San Francisco (Petrich et al. 2013). At the Tinkering Studio, the adults responsible for facilitating learning (museum floor staff) support student learning by "**Spark[ing]** initial interest and participation through demonstrations, modeling, parallel play or questions; **Sustain[ing]** participation through frustration, distraction or boredom through introduction of new tools, approaches or analysis; and **Deepen[ing]** understanding and commitment through *complexification* of concepts and making connections to learners' interests" (Gutwill et al. 2015). The description of the role of floor staff at the MAKESHOP is similar.

The facilitation guidelines outlined by the museum maker spaces were useful, but the course instructor also incorporated more general ideas from informal science education and museum research. During the full time of the study, the instructor was collaborating on the design of the exhibits at a new museum, the Wolf Museum of Exploration + Innovation (MOXI), and that work on informal science education influenced her perspectives and course activities. MOXI's exhibits were being planned as the NGSS were authored, providing an opportunity to design a museum fully informed by the NGSS and a place for pre-service teachers to observe children learning NGSS-aligned content in an informal science space (Harlow et al. 2017). In the pre-service teacher education course studied here, teacher candidates read and discussed the documents about facilitating learning in these informal spaces, highlighting strategies they found particularly helpful or promising for their own future instruction and then visited MOXI to observe children's interactions in informal spaces.

Thus far, we have only been discussing how teachers facilitate interaction. However, in classrooms, teachers are held accountable by state and district assessment standards. Assessing maker education activities is particularly challenging, especially when the goal is to document learning without interrupting the joy and playfulness of the maker activity. Like the NGSS, researchers in maker education highlight the importance not only of the declarative knowledge that participants develop but also *how* learners are "engaging in practices that draw on facts and skills to advance valued and purposeful activity" (Petrich et al. 2013, p. 69). Blikstein and Worsley (2016) emphasized the need to change the culture surrounding assessment to prioritize *process over product*. Focusing assessment and attention only on the finished product can discourage students from taking risks and pursuing more complex design ideas in fear of failure. Instead, maker educators recommend process-based assessments in which students document their design decisions over time.

Of the specific strategies for assessing educative making, portfolios are particularly promising because they can provide rich evidence of student learning.

In particular, open portfolio systems in which the maker decides what to include have been studied for their potential to document educative making projects (Litts et al. 2016). Unlike traditional assessment and documentation tools, portfolios are learning tools, allowing students to present both their process, including failed attempts and intermediate steps, and their final product and to revisit their own decisions. Teachers should provide ongoing and specific feedback to students on "works-in-progress," being careful to only praise students for things they can change about a project or design (Regalla 2016) in order to encourage a growth mind-set (Dweck and Leggett 1988). Portfolios can help teachers provide timely feedback while also documenting students' progress over time. In fact, some colleges and universities now accept maker portfolios as part of students' applications. In the course studied here, the pre-service teachers were encouraged to develop assessment tasks that did not distract from the maker activity, which could include portfolios, or informal interview questions, design drawings, or using artifacts as evidence of learning.

14.3.3 Materials Knowledge: Knowledge of Tools and Materials

The final area, "materials knowledge" may be the most often associated with educative making. Making is characterized by the products people create to be used, displayed, or shared within and across communities (Martin 2015). Considering the focus on creation, tools play a key role in defining the learning environment, activities, and end products. When advising teachers new to maker activities, tools are among the first things mentioned and are considered crucial to implementing activities and fostering a maker ethos of creative problem-solving (Smith and Smith 2016).

Maker activities often (but not always) include digital or electronic fabrication, and making is associated with tools such as 3D printers, laser cutters, and circuits. These new technologies may be integrated into traditional crafts (Honey and Kanter 2013; Martin 2015; Peppler et al. 2016). Activities can include "low-tech" materials such as cardboard or pipe cleaners and benefit from traditional tools such as sewing machines and drills, but "maker spaces incite our imagination because they contain new objects as well, such as 3D printers, CNC mills, and hobbyist electronics, creating environments where you can design virtually anything" (Peppler et al. 2016, p. 4).

Despite their promise to spark curiosity, these materials and tools alone will not lead to learning. Resnick and Silverman (2005), in designing maker tools, have emphasized the importance of "low floors, high ceilings, and wide walls," (p. 118) where novice learners have an easy entry point but can develop deeper understanding with more difficult projects and options to follow their interests. The same tools and materials can present diverse content and reach students of varying abilities (Alper 2013), but activities must be thoughtfully designed and implemented to encourage deep rather than superficial learning (Blikstein and Worsley 2016).

Teachers must understand that some materials are more likely to lead students to "bump up" against the desired content knowledge (Bennet and Monohan 2013). These materials lead to students solving problems while developing sophistication at the intersection of content knowledge and materials knowledge. Educators must also understand that the tools only hold potential if they are appropriate for the cognitive and physical capabilities of the children (intersection of pedagogical knowledge and materials knowledge). Regardless of the differences in tools provided, students benefit from learning how to select appropriate tools, understanding the uses and limitations of tools, and learning hands-on approaches to problem-solving and design (Leak et al. 2018).

14.4 Example Learning Activities

Following, we provide an overview of the course learning experiences and assignments used to support pre-service elementary school teachers in their efforts to develop, facilitate, and assess an educative making activity. We highlight selected learning experiences from the course in more detail to provide additional context and recommendations for teacher educators. Table 14.1 includes the topics in the most recent year of the science methods course that relate directly to helping teacher candidates develop MPCK. While there was at least one such activity each week, these do not constitute all the course activities. Activities not related to maker education are not discussed here. Activities highlighted with an asterisk are discussed in more depth in this section.

14.4.1 Week 1: Documenting Learning – NGSS and Portfolios

On the first day of class, after a brief overview of NGSS, students were introduced to educative making through a shadow puppet activity. They created a short video demonstrating a science core idea and started an online digital portfolio. While the students were free to choose any area of science to create their video about, the lesson was tied to the disciplinary core ideas around light and shadow. Following instructions and a video on learnxdesign.org (a resource developed by a consortium of science and technology museums), the teacher candidates designed storyboards and then created a video using shadow puppets, which they uploaded to the digital portfolio system. The pre-service teachers observed and analyzed each other's videos for indicators of learning. Finally, the pre-service teachers worked in groups to collaboratively design an assessment around the NGSS standards about light and shadows. At this point, most of the assessments they designed consisted of paper and pencil tests, even though they had just described how their videos demonstrated competency (Fig. 14.2).

Table 14.1 Overview of course activities related to maker education (*indicates weeks described in more detail in sections 14.4.1–14.4.5)

Wk	Topic	Content	Pedagogy	Materials
1*	Making and documenting learning: NGSS and portfolios	Light and shadow	Electronic portfolios	Cardboard, sticks, scissors, video cameras, lights
2	Understanding the disciplines of science vs. engineering	Science vs. engineering design (NGSS practices)	Engaging students in the practices of science and engineering	Cardboard, crafting materials
3*	Understanding learning progressions: An example through circuits	Electricity	Learning progressions	Paper circuits, squishy circuits, simple circuits
4	NGSS alignment: Analyzing lessons and curriculum	Earth science, engineering	Materials setup, facilitating initial engagement	Magnets, 3D printer
5 and 6*	Attending to ideas: Indicators of learning from informal spaces Field trip to museum	Physical science	Facilitating engagement, field trips	Many different materials found at museum
7	Building and coding robots	Engineering, computer coding	Integrating content areas	Lego robotics, scratch programming
8	Mock maker faire	Putting all areas into practice. Trying out activities with peers. Redesigning		
9*	School maker faire	Putting all areas into practice. Working with children		

Fig. 14.2 Making shadow puppet videos

Fig. 14.3 From left to right: Squishy circuits, paper circuits, Makey Makey circuits

14.4.2   Week 3: Understanding Learning Progressions – An Example Through Circuits

During the third week of class, we introduced the idea of learning progressions (NRC 2007), building on student ideas and electric components. To accomplish this, we divided the class into three groups based on the grade level of their assigned practicum classroom (K–1, 2–4, 5–6). All students did a pre-assessment activity in which they predicted which of nine arrangements of batteries, bulbs, and wires would result in a lit bulb. They then tested these arrangements with actual batteries, bulbs, and wires and then discussed rules that seemed to apply to all the arrangements that resulted in a lit bulb. The K–2 group then played with squishy circuits (Johnson and Thomas 2010), the 3–4 group made Valentine's Day cards from paper circuits, and the 5–6 group used Makey Makey to create game controllers and edited scratch game programs that were compatible with the Makey Makey. The pre-service teachers documented their own decision-making through this process with the electronic portfolio application and then considered what and how they would assess NGSS standards for their grade level (Fig. 14.3).

14.4.3   Weeks 5 and 6: Indicators of Learning from Informal Spaces and Museum Field Trip

During week 5, teacher candidates were presented with informal indicators of learning designed by science museums with a focus on maker education. In class, students read about the pedagogical (facilitation and design) strategies developed by the Tinkering Studio and MAKESHOP. They discussed in class how science they saw in their practicum classrooms was different or similar to science in museums. Then they designed their own learning dimensions and facilitation strategies to use in their future instruction.

The following week, the pre-service teacher candidates went on a field trip to MOXI, a local science museum that focuses on engaging children in physical science concepts through play-based, interactive exhibits. During this trip, teacher candidates were prompted to observe children engaging with the exhibits, as well as design a complementary activity to build on or extend learning in conjunction with one or more exhibits at the museum.

14.4.4   Additional Maker Activities

Throughout the course, pre-service teacher candidates engaged in other educative making activities to further refine their developing understandings related to the MPCK model. For example, they created cardboard automata following the Exploratorium's Guide, built and programmed robots, and explored foldable microscopes to combine art and science. One week, they completed an abbreviated version of an online 3D printing challenge to develop technology that could help astronauts prepare food. From this challenge, they learned about engineering design, Tinkercad, and 3D printing, along with science content related to space science and life science. Reflections following the activities focused on connections to the NGSS and how to assess whether or not students were learning without disrupting the maker experience.

14.4.5   Week 9: School Maker Faire

The School Maker Faire was an event for the community scheduled during the pre-service teachers' regular class meeting time during the second to last week of the course. It was an opportunity to showcase and practice what they had been working on. At the School Maker Faire, the pre-service elementary school teachers led activities alongside informal science education providers and university outreach programs. The stations were set up in five university classrooms and lawn spaces around the education building where the pre-service teachers took classes for their credential program. The children in the kindergarten through sixth grade classrooms that the pre-service teachers were placed in were invited to attend with their families. Once at the event, children were free to wander with their families to any station that intrigued them and stay for as much or as little time as they wanted. This required pre-service teachers to design and facilitate activities in a way that appealed to student interests. Also, because the event included families, young children, older children, and adults who were often all present at a station simultaneously, pre-service teachers were required to plan for multiple entry points to their activities. The activities designed by the pre-service teachers included making slime, cardboard pinball machines, apple boats, and kaleidoscopes (see Fig. 14.4). All were explicitly connected to NGSS standards.

Fig. 14.4   From left to right: Kaleidoscopes, apple boats, cardboard pinball

14.4.6   Reflections

After the event, the pre-service teachers reflected on their experiences. In an online reflective survey, we asked them open-ended questions about the experience, children's learning, and their own learning. In response to a question that asked about whether anything surprised them about the children that interacted with their station, a majority of the pre-service teachers mentioned the interest of children. Pre-service teachers reported they were surprised how long children stayed focused at their stations and their interest in figuring out how to make things, regardless of grade level. For example, one teacher candidate commented, "I was surprised at how invested they were in designing and constructing their model and accommodating their marble run to fit the challenge," and another stated "I was surprised at the versatility. We had 4-year-olds successfully participating (with some help) as well as some 12-year-olds who talked about aerodynamics and made more creative designs." When asked what they themselves learned by facilitating the activity, pre-service teachers commented on learning that children enjoyed making things and solving problems and that the children were more creative than they expected. Nearly all the pre-service teachers commented on the fun that they and the children had, "This was such a fun way to teach and learn!"

14.5   Conclusion

The knowledge required to design, facilitate, and assess educative making activities lies at the nexus of pedagogical knowledge, content knowledge, and knowledge of materials. The MPCK model proposed is more complex than either science PCK or engineering PCK and provides insight into implementation strategies in the new domain of educative making.

Teachers can incorporate educative making into thematic units ranging from electricity to earth science and simultaneously include engineering design practices such as defining a problem or optimizing the design solution. In addition to content knowledge, educators need pedagogical knowledge to initiate and sustain engagement. Educators also need to rethink documentation and assessment in ways that value process over product and do not detract from the flow and enjoyment of making, such as interviews, design drawings, or portfolios. Finally, educators need materials knowledge to help students understand the uses and limitations of tools and hands-on approaches to problem-solving and design. Teacher education programs targeting the three knowledge areas of the MPCK model can increase new teachers' competence and comfort levels and provide new ways to engage and excite learning in science and engineering.

Helping pre-service teachers develop such knowledge requires careful structuring. Further, while we focused on the *knowledge* base, engaging children in educative making also requires teachers to have attitudes toward learning that are aligned with educative making, confidence in their ability to facilitate student learning through such activities, and a network of supportive administrators and other teachers. Practicing teachers will also benefit from community partners such as museums, universities, and local businesses who can provide expertise and access to tools and materials. Through such partnerships and the new model of pedagogy and teacher development introduced in this paper, maker-inspired education and its benefits for learning science can become accessible to all students. As Dewey stated over a century ago, "The instinct of making—the constructive impulse. The child's impulse to do finds expression first in play, in movement, in gesture, and make-believe, becomes more definite and seeks outlet in shaping materials into tangible forms and permanent embodiment" (Dewey, 1915/2001, p. 30).

References

Alper, M. (2013). Making space in the makerspace: Building a mixed-ability maker culture. In Proceedings of the 2013 Conference on Interaction Design and Children. New York: ACM.

Bennet, D., & Monohan, P. (2013). NySci design lab. In M. Honey & D. Kanter (Eds.), *Design make play: Growing the next generation of STEM innovators*. New York: Routledge.

Bevan, B. (2017). The promise and the promises of making in science education. *Studies in Science Education, 53*, 75. https://doi.org/10.1080/03057267.2016.1275380.

Blikstein, P., Chen, V., & Martin, A. (2014). Promoting diversity within the maker movement in schools: New assessments and preliminary results. In *Proceedings from ICLS '14: International Conference of the Learning Sciences*. Boulder, CO: International Conference of the Learning Sciences.

Blikstein, P., & Worsley, M. (2016). Children are not hackers: Building a culture of powerful ideas, deep learning, and equity in the Maker Movement. In K. Peppler, E. R. Halverson, & Y. B. Kafai (Eds.), *Makeology: Makerspaces as learning environments* (Vol. 1, pp. 64–79). New York: Routledge.

Cantrill, C., & Oh, P. (2016). The composition of making. In K. Peppler, E. Halverson, & Y. B. Kafai (Eds.), *Makeology: Makerspaces as learning environments* (pp. 107–120). New York, NY: Routledge.

Clough, G. W. (2004). *The engineer of 2020: Visions of engineering in the new century*. Washington, DC: National Academy of Engineering. https://doi.org/10.17226/10999.

Dewey, J. (1915/2001). *The school and society* (p. 30). Minola, NY: Dover Publications.

Dorph, R., Sheilds, P., Tiffany-Morales, J., Hartry, A., & McCaffrey, T. (2011). *High hopes-few opportunities: The status of elementary science education in California*. Sacramento, CA: The Center for the Future of rTeaching and Learning at WestEd.

Dweck, C., & Leggett, E. L. (1988). A social-cognitive approach to motivation and personality. *Psychological Review, 95*(2), 256–273.

Fosnot, C. T. (2013). *Constructivism: Theory, perspectives, and practice*. New York: Teachers College Press.

Fosnot, C. T., & Perry, R. S. (1996). Constructivism: A psychological theory of learning. In C. T. Fosnot (Ed.), *Constructivism: Theory, perspectives, and practice* (pp. 8–33, 2nd ed.). New York: Teachers College Press.

Garneli, B., Giannakos, M. N., Chorianopoulos, K., & Jaccheri, L. (2013). Learning by playing and learning by Making. In *International Conference on Serious Games Development and Applications* (pp. 76–85). Berlin Heidelberg: Springer.

Gay, G. (2010). *Culturally responsive teaching: Theory, research, and practice*. New York: Teachers College Press.

Gutwill, J. P., Hido, N., & Sindorf, L. (2015). Research to practice: Observing learning in tinkering activities. *Curator: The Museum Journal, 58*(2), 151–168.

Harlow, D., & Hansen, A. (2018). School maker faire as pre-service teacher education. *Science & Children, 55*(7), 30–37.

Harlow, D., Skinner, R., & O'Brien, S. (2017). Roll It Wall: Developing a framework for evaluating practices of learning. In *Proceedings of the 7th Annual Conference on Creativity and Fabrication in Education*. https://doi.org/10.1145/3141798.3141813.

Honey, M., & Kanter, D. E. (Eds.). (2013). *Design, make, play: Growing the next generation of STEM innovators*. New York, NY: Routledge.

Johnson, S., & Thomas, A. P. (2010, April). Squishy circuits: A tangible medium for electronics education. In *CHI'10 Extended Abstracts on Human Factors in Computing Systems* (pp. 4099–4104). New York, NY: ACM.

Koehler, M. J., & Mishra, P. (2008). Introducing TPCK. In *Handbook of technological pedagogical content knowledge (TPCK) for educators* (pp. 3–29). Mahwah, NJ: Lawrence Erlbaum Associates.

Leak, A. E., Santos, Z., Reiter, E., Zwickl, B. M., & Martin, K. N. (2018). Hidden factors that influence success in the optics workforce. *Physical Review Physics Education Research, 14*(1), 010136.

Litts, B., Kafai, Y., Fields, D., Halverson, E., Peppler, K., Keune, A., Tissenbaum, A., Grimes, S., Chang, S., Regalia, L., Telhan, O., & Tan, M. (2016). *Connected making: Designing for youth learning in online maker communities in and out of schools*. 12th International Conference of the Learning Sciences (ICLS 2016) Singapore (20–24).

Maltese, A. V., & Tai, R. H. (2011). Pipeline persistence: Examining the association of educational experiences with earned degrees in STEM among US students. *Science Education, 95*(5), 877–907.

Martin, L. (2015). The promise of the maker movement for education. *J-PEER, 5*(1). Article 4.https://doi.org/10.7771/2157-9288.1099.

Martinez, S. L., & Stager, G. (2013). *Invent to learn: Making, tinkering, and engineering in the classroom*. Torrance, CA: Constructing modern knowledge press.

Mulvey, P. & Pold, J. (2015). *Physics Bachelor's Initial Employment, Tech. Rep.* College Park, MD: American Institute of Physics.

National Research Council. (2007). *Taking science to school: Learning and teaching science in grades K-8*. Washinigton, DC: National Academies Press.

National Research Council. (2012). *A framework for K-12 science education: Practices, crosscutting concepts, and core ideas*. Washington, DC: National Academies Press.

National Science Foundation & National Center for Science and Engineering Statistics. (2015). *Women, minorities, and persons with disabilities in science and engineering: 2015*. Special Report NSF 15-311. Arlington, VA. Available at http://www.nsf.gov/statistics/wmpd/

NGSS Lead States. (2013). *Next Generation Science Standards: For states, by states*. Washington, DC: National Academies Press.

Papert, S. (1980). Mindstorms: Children, computers, and powerful ideas. New York: Basic Books.

Papert, S., & Harel, I. (1991). Situating constructionism. *Construction, 36*(2), 1–11.

Peppler, K. (2016). ReMaking arts education through physical computing. In K. Peppler & E. H. Y. B. Kafai (Eds.), *Makeology: Makers as learners* (pp. 206–226). New York, NY: Routledge.

Peppler, K., & Bender, S. (2013). Maker movement spreads innovation one project at a time. *Phi Beta Kappan, 95*(3), 22–27.

Peppler, K., Halverson, E., & Kafai, Y. B. (Eds.). (2016). *Makeology: Makerspaces as learning environments* (Vol. 1). New York, NY: Routledge.

Petrich, M., Wilkinson, K., & Bevan, B. (2013). It looks like fun, but are they learning? In Honey, M., & Kanter, D. E. (Eds.), Design. Make. Play. Growing the next generation of STEM innovators (pp. 50–70). New York, NY: Routledge.

Quinn, H., & Bell, P. (2013). How designing, making, and playing relate to the learning goals of K-12 science education. In M. Honey & D. Kanter (Eds.), *Design make play: Growing the next generation of STEM innovators* (pp. 17–33). New York, NY: Routledge.

Regalla, L. (2016). Developing a maker mindset. In K. Peppler & E. H. Y. B. Kafai (Eds.), *Makeology: Makerspaces as learning environments* (pp. 257–272). New York, NY: Routledge.

Resnick, M., & Silverman, B. (2005). Some reflections on designing construction kits for kids. In *IDC'05 Proceedings of the 4th International Conference on Interaction Design and Children* (pp. 117–122). New York: ACM.

Sawyer, R. K. (2004). Creative teaching: Collaborative discussion as disciplined improvisation. *Educational Researcher, 33*(2), 12–20.

Shulman, L. S. (1987). Knowledge and teaching: Foundations of the new reform. *Harvard Educational Review, 57*(1), 1–22.

Simpkins, S. D., Davis-Kean, P. E., & Eccles, J. S. (2006). Math and science motivation: A longitudinal examination of the links between choices and beliefs. *Developmental Psychology, 42*(1), 70.

Smith, W., & Smith, B. (2016). Bringing the Maker Movement to school. *Science & Children, 54*(1), 30–37.

Tai, R., Liu, C., Maltese, A., & Fan, X. (2006). Planning early for careers in science. *Science, 312*, 1143. https://doi.org/10.1126/science.1128690.

Vossoughi, S., Hooper, P. K., & Escudé, M. (2016). Making through the lens of culture and power: Toward transformative visions for educational equity. *Harvard Educational Review, 86*(2), 206–232.

Wardrip, P. S., & Brahms, L. (2015, June). Learning practices of making: developing a framework for design. In *Proceedings of the 14th international conference on interaction design and children* (pp. 375–378). New York, NY: ACM.

Wardrip, P. S., & Brahms, L. (2016). Taking making to school: A model for integrating making into classrooms. In K. Peppler & E. H. Y. B. Kafai (Eds.), *Makeology: Makerspaces as learning environments* (pp. 97–106). New York, NY: Routledge.

Index

Printed in the United States
By Bookmasters